Toward a Critical Instructional Design

Edited by Jerod Quinn, Martha Fay Burtis, and Surita Jhangiani
Foreword by Robin DeRosa

Hybrid Pedagogy Inc.
2022

© 2022 Hybrid Pedagogy
Chapters © their respective authors

Published by Hybrid Pedagogy Inc.
https://hybridpedagogy.org
Denver, CO

Cover art by Pawel Czerwinski
Book design by Martha Fay Burtis

Toward a Critical Instructional Design by Hybrid Pedagogy Inc. is licensed under a Creative Commons Attribution-NonCommercial 4.0 International License, except where otherwise noted.

First printing, August 2022
ISBN 979-8-9866764-1-8

Contents

Acknowledgements — v

Foreword — vii
Robin DeRosa

Introduction — xi
Martha Fay Burtis and Jerod Quinn

Counter-Friction to Stop the Machine — 1
Jesse Stommel and Martha Fay Burtis

Centering the Margins Through Critical Instructional Design — 13
Amy Collier, Sarah Lohnes Watulak, and Jeni Henrickson

Quiddity of Design in [learning + design] — 31
Nicola Parkin

Compassionate Learning Design as a Critical Approach to Instructional Design — 49
Daniela Gachago, Maha Bali, and Nicola Pallitt

Indigenizing Design for Online Learning in Indigenous Teacher Education — 73
Johanna Sam, Jan Hare, Cynthia Nicol, and LeAnne Petherick

Hybrid Teaching is Not a Limbo Nor a Multiverse — 95
Victor Azuaje

Building a Framework — 107
Mary Mathis Burnett

On Practice: Instructional Designer — 125
Autumm Caines

Quality Theater
Jerod Quinn
127

Siberian Syndrome in Online Learning
Natalie Shaw
149

Building from Our Sensitive Edges
Meryl Krieger and Clayton D. Colmon
163

Militaristic Origins, Power, and Faux-Neutrality
Hannah Hounsell
189

Towards a Critical Instructional Design Framework
Katrina Wehr
201

Author Biographies
217

Acknowledgements

The editors of this collection would like to thank, first and foremost, the authors who so graciously shared their work with us for this collection. Thank you for staying with us through the call for chapters, months of editing and revisions, a change in editorial lineup, and the long and drawn out process that is book creation. We are deeply proud of this collection, and we hope you are too.

We would also like to specifically thank Sean Michael Morris for igniting the community around critical instructional design, and for seeing a space for this conversation in putting out the call for chapters through Hybrid Pedagogy. Thank you also for your help selecting chapters, for those first rounds of edits, and for handing over this project to us and trusting us to see it through to the end.

Thank you to Jesse Stommel as well, for your production guidance and all the wrangling of Pressbooks and the publication process. We appreciate the help and editing advice you have offered, along with your time and energy on this project. Also, the Zoom cameos from Hazel are always a welcome addition to any meeting.

Thank you, Sukaina Walji, for your help in selecting the chapters for this collection along with the early rounds of edits and revisions. Your assistance and expertise were greatly appreciated.

Jerod would like to also thank the following amazing people for their help, willingness to listen, high-fives, and general encouragement: Amy Archambault, Amanda Stafford, Fatemeh Mardi, Tammy M. McCoy, Chris Grabau, Allen Brown, Donohon Abdugafurova, Norah Elmagraby, and Matt Easter. Thank you to my family as well, especially Charity, who spent many nights helping me think through the dynamics and details of book editing along with helping me to find constructive yet kind ways to say, "I need you to cut 1000 words from this chapter because it's too freakin' long." And while my kids didn't have much to do with this collection, they like seeing their names in books so here you go Grace and Ash.

Surita would like to thank Jerod and Martha for being such amazing editors to work alongside with. I am grateful for the opportunity to have been a part of this project. I learned so much along the way about unique approaches and issues related to critical design and care.

Martha would like to thank Robin, Hannah, and Matt for helping her survive the last three years; Susan and John Fay for being the first people to teach her why writing (and editing!) matters; and Erik, Madigan, and Graeme for being the reasons everything else matters.

Foreword

Robin DeRosa

Indulge me while I tell a little story about a hypothetical faculty member. Let's call her "Robin." Robin accepted an adjunct teaching position at a rural public college in the Northeast United States as she was finishing her doctoral dissertation. As an early Americanist studying at a prestigious R1 university, Robin was extremely expert in several narrow areas related to 17th-century literature and history, which was disappointingly unhelpful to her as she waded into her first assigned courses: "Composition" and "Modern British Literature" (someone was on sabbatical, which still doesn't explain why anyone thought she was qualified for that last one). In fact, even a year later when she was able to begin teaching "in her field" as a one-year full-time contracted faculty member, Robin realized that her main strategy for course design revolved around adapting 1) the syllabi of her grad school professors (many of them brilliant researchers with no pedagogical training and often very spotty abilities to interact pleasantly with other humans) and 2) the activities of her middle school teachers (particularly Miss McNeil, whose 8th-grade social studies unit on seagull regurgitation is fondly remembered by countless cohorts of Massachusetts public school graduates). When Robin was privileged and lucky enough to land a tenure-track position in English, she realized that she had almost no formal preparation for teaching the four courses per semester—most outside of her specific field—that she was assigned. While scholarship and service were now part of her professional expectations, most of her work hours were focused on teaching, a profession for which she was wholly untrained.

I may as well stop pretending this is a hypothetical. I should also stop pretending it's generalizable, since any higher ed anecdote that contains the phrase "land a tenure-track job" is already far too fantastical these days. But my story is likely a familiar one for many college-level instructors. With the exception of those who teach education, most of us secure our teaching positions primarily by demonstrating content expertise. Once we join a university faculty, we may have access to professional development workshops run by a teaching and learning center, and we may be lucky enough to have instructional designers to partner with when we need help rethinking an assignment, tweaking a syllabus, or setting up effective discussions in our learning management system. Tenure-track professors at R1 institutions may actually

Toward a Critical Instructional Design

be disincentivized from focusing on teaching, as it can stall their progress towards tenure and promotion by taking their time away from research. Meanwhile, contingent faculty may be balancing high course loads and/or have no access to funding or compensation for time spent in professional development, and community college faculty may find their institutions lack teaching centers or instructional support staff. Even as we realize through our years of teaching that we, as faculty, are both in need of and capable of improving our teaching, higher ed as a whole simply does not prioritize the development of teaching faculty.

For many years, I didn't really understand how problematic this was for my own work. After all, I was committed to a "student-centered" classroom, genuinely invested in my students and their learning processes, and open to experimenting with new technologies and tricks. Working sporadically with instructional designers and academic technologists was eye-opening, as it dawned on me that people had actual training in the kinds of architectures that underpin a course, architectures that I was certainly using, but which were mostly invisible to me as I remained focused on my all-important content.

More eye-opening, though, was watching the edges of higher education begin to fray, and indeed my own public university begin to unravel as the terrain of higher ed shifted beneath our feet. This ongoing fraying is less a random upheaval than it is a series of revelations about deep-seated problems in how our institutions are structured coming to violent fruition. Some areas of the United States where I live and teach, for example, are wrestling with demographic trends that are eroding college enrollments; combine that with the defunding of public higher education, a massive student loan debt crisis, and a national rhetoric that (given the costs that must be borne by individual families) questions the value of college, and you have a cocktail that has caused many institutions to become austerity-drunk—cutting services, supports, and staffing to the bone. Many states, including my own, have passed "divisive concept" laws preventing teachers at many levels from engaging honestly with concepts such as systemic racism, LGBTQ+ history, and genocide. And COVID has pushed every institution to its limits with bills to pay for testing and PPE; challenges keeping students, staff, and faculty safe in campus settings; and pivoting to online modalities without advance preparation.

On the one hand, we have many students struggling: impoverished and unable to afford food if they purchase the required books for their classes; immunocompromised and afraid to return to class now that mask mandates are dropping and online access is being revoked as an option; suffering deeply from prejudice and discrimination that makes it far less likely for BIPOC students at predominantly white institutions to graduate; suffering from mental illness as counseling centers are overrun; navigating neurodivergence as accessibility services are underresourced; and facing pressure from families to enroll in majors that they don't enjoy based on often-flawed speculation about which fields will deliver a much-needed return on investment.

On the other hand, we have institutions focused not on these alarming struggles, but on their own. We see the most elite institutions battling each other to prove just how selective they are; in other words, they gain prestige as fewer and fewer students who want to attend are able to do so. At the same time, our open access institutions, our regional public institutions, our community colleges, our minority-serving institutions and HBCUs—the places that actually educate the vast majority of students—struggle, literally, to keep the lights on. And instead of leaning into the equity challenges that have come to a boiling point, they lean into fiscally-focused "solutions" that do nothing to relieve the heat.

If you don't teach at an elite institution (and most of us don't!), I bet you've felt the burn of these solutions. Has your institution rebranded or spent money on marketing while claiming not to have funding to support expanded instruction or academic services? rallied around majors that are considered "marketable" while cutting majors that are not directly tied to workforce demand? reduced tenure-track faculty and expanded use of adjuncts or "professors of practice?" outsourced to Online Program Management companies (OPMs) to quickly offer high-demand online majors or grad programs? consolidated departments or even whole schools? increased discount rates to undercut competitor institutions and attract enrollments? And if your institution has done any of these things, has that successfully relieved the budgetary pressures on your institution? I am guessing no. And what has been the effect of these things on your struggling students? I can guess that as well.

So while I would argue that it is important that higher ed faculty be offered support and professional development as they embark on teaching careers for which they are generally under-prepared, it's important to see this support and professional development as part of a critical intervention that speaks to the confluence of storms that are buffeting us. Which brings me to my gratitude for this collection.

"Critical Instructional Design," as it will take shape in this volume, advocates for a reframing of how we think about the supports that faculty need to improve their course designs. Instead of an outsourcing model, where support staff (or external companies!) provide the design expertise, CID seeks to connect designers, technologists, learning center staff, librarians, and other key pedagogical partners with faculty to help faculty learn what they need in order to make intentional decisions about their course architectures. And unlike de-contextualized and standardized checklists or rubrics, CID seeks to situate course design in time and space, encouraging faculty to think of their specific students and their needs, their specific institutions and their missions, and the specific political and social contexts that surround their courses each time they teach them. What you will find in this volume will be less a roadmap to best practices in instructional design, and more an invitation into a learning community that trusts faculty to engage thoughtfully with the complexities of course design.

We are all keenly aware that the world is currently facing severe crises, and that many of these are changing the shape of our colleges and universities—and perhaps even the shape of learning itself. It's time for an instructional design that brings the full power of faculty to bear on the challenges that we face. As we design our courses, we are participating in a larger project to design the future of higher education, and as we develop our skills in teaching, we must also develop our engagement with the larger questions of what our students will need to thrive in a world that is currently presenting so many challenges to their livelihood, health, and happiness. I hope you will engage with this book not just as a catalyst for creative inspiration about how to approach course design, but also as a call to action for how our grassroots work teaching and learning with our students can open hopeful and humane futures for both higher education and for our world.

Introduction

Martha Fay Burtis and Jerod Quinn

As we all witnessed our schools scrambling to deal with the COVID-19 pandemic, the implications of instructional design had never felt so important—or so fraught. As our schools now scramble to "return to normal," we wonder what we've learned? For many of us, the crisis brought into focus that the assumptions underpinning our instructional design, and our ID practices themselves, were failing our teachers, students, and institutions. Perhaps it was time for a new approach?

In March of 2021, Hybrid Pedagogy put out a call for chapters for a Critical Instructional Design Reader. That call asked the questions:

> "What if technology had misled us, distracted us from what's actually important for teaching online? What if technology has so far interpreted instruction for us—even from the days of correspondence courses—making the page, digital or otherwise, a surrogate for our pedagogies? How do we reclaim the relational, communal, intimate side of teaching when glass and pixels and apps stand between? When we undertake the work of defining and investigating critical instructional design, we must shift our focus from the screen to the student, from best practices to humanizing pedagogies."

Submissions from North America, Europe, Africa, and Australia came in with a wide scope for how and why a problem-posing approach of critical pedagogy can be applied to online classes. We heard from instructional designers, educators, and students themselves. It quickly also became clear that while the COVID-19 global pandemic did not instigate these conversations, it certainly poured gasoline on the fire of implementing them. Care in online classes began to become a mainstream conversation among all kinds of educators as the pandemic created new tensions and exacerbated old ones on a literal global scale.

As the editors were sifting through the submissions we noticed two related, but divergent streams emerging. One stream was focused on creating environments and experiences grounded in care and compassion. It includes conversations about the logistical hurdles of building intentional hospitality into online experiences, but also the rewards of learners being included in the experience. The other stream focused on applying the ideals of critical instructional design to the course design process. These chapters challenged the assumptions of linear, western approaches to higher education and pushed the boundaries of what online learning can look like. The editors made the decision to gather these chapters into two sibling collections: Designing for Care and Toward a Critical Instructional Design. The collection you are reading is the Toward a Critical Instructional Design edited collection.

Toward a Critical Instructional Design

The idea of critical instructional design, while it is still nascent and demands continued commitment from us as we strive to define it and enact it, is rooted in the ideas of critical pedagogy first laid out by Pablo Freire. In *Pedagogy of the Oppressed*, he famously compared traditional education to banking:

> "In the banking concept of education, knowledge is a gift bestowed by those who consider themselves knowledgeable upon those whom they consider to know nothing...The students, alienated like the slave in the Hegelian dialectic, accept their ignorance as justifying the teachers existence — but unlike the slave, they never discover that they educate the teacher."

Freire's critique of the banking model of education is a cornerstone of critical pedagogy. And through that metaphor, he not only illustrates for us what is wrong with the status quo, he also begins to draw a path for us in another direction, leading us toward a problem-posing pedagogy:

> "Those truly committed to liberation must reject the banking concept in its entirety, adopting instead a concept of women and men as conscious beings, and consciousness as consciousness intent upon the world. They must abandon the educational goal of deposit-making and replace it with the posing of the problems of human beings in their relations with the world."

In the same way that Freire imagined a metaphor of capital (and its implied hoarding and distribution) as a metaphor for teaching, we would do well to understand the metaphors (and their implied practices) that underpin traditional instructional design. In fact, the roots of the field, in military and corporate training, point us in useful directions. Traditional ID often seems consumed with a militaristic and

Introduction

industrial ethos—and linear, mechanistic ways of describing learning (and, by extension, learners). It is demarcated by structures, rubrics, tools, and checklists. It fusses over design rules in which objectives are tidily mapped to activities that are neatly measured by assessments. In all this fussiness and tidiness, mechanics and measurements, it loses sight of the complex and messy humans that sit at its center—human teachers, students, and instructional designers. Freire describes the banking model of education as alienating and enslaving; similarly, traditional instructional design approaches separate learners from their humanity, expecting them to act as yet another (predictable and impersonal) cog in the industrial machine of learning.

Just as Freire insisted that we center humans in pedagogy, critical instructional design demands we find a way to center humans in the design of education. We must learn to live without our fussy tools and find new ways of imaging and describing the work of instructional design. What can we imagine, design, create, build that will liberate our design practices—and the humans inside of them—the way Freire helped liberate our classrooms—and the teachers and students inhabiting them? We hope that this volume is just the first exchange in a new and sustained conversation about critical instructional design.

Scope of the Collection

In this collection, you will find perspectives from across the globe engaged in a conversation about the purpose, meaning, and practice of critical instructional design. We are particularly proud of the wide range of voices included here, from students to designers to technologists to teachers. In addition, we have authors from Australia, Africa, Europe, and North America all sharing their work. We decided that changing the spelling of words to be the Americanized versions seemed to work against the value of letting people share their own stories in their own words. It might be a "little" thing, but it didn't feel right to change these so you will see spelling variants throughout the collection. That choice was intentional.

Counter-Friction to Stop the Machine starts this collection by describing how current approaches undermine teachers, devalue instructional designers, and harm students. In the shadow of the COVID-19 pandemic, more and more schools are doubling down on these problematic practices, making the need for a new, critical approach to instructional design all the more acute.

Centering the Margins through Critical Instructional Design situates us directly in the long shadow that COVID-19 has cast over colleges and universities. Importantly, they place the real, lived experiences of students at the center of the need for a critical instructional design, and they share specific practices to counter the harm students face.

Quiddity of Design in [learning + design] proposes we think more deeply about the philosophical dimensions of learning design. Through this probing into the "whatness" of our work we can learn to better understand, and operate within, the tensions, discomforts, and risks that often make their home in education.

Compassionate Learning Design as a Critical Approach to Instructional Design suggests we embrace compassion in our teaching and introduces a model for compassionate learning design that is structured around participation, justice, care, and an overarching commitment to praxis.

Indigenizing Design for Online Learning in Indigenous Teacher Education demonstrates how we can embrace and embed Indigenous perspectives and histories in our course designs. In particular, the authors focus on four pedagogical principles informed by Indigenous experiences: Indigenous knowledge frameworks, localization, multimodalities, and designing for relationships.

Hybrid Teaching is Not a Limbo Nor a Multiverse explores the HyFlex phenomenon that has recently permeated the higher education landscape and identifies one of its core challenges: teachers cannot sustain the "double-identity model" that is so often attempted. Alternatively, by adopting a Freirien dialogical approach, educators can develop a consistent teaching persona across modalities and contexts.

Building a Framework shares the story behind the development of an Application Framework for Critical Pedagogy. As a tool, the framework was designed to help faculty across a spectrum of experience to engage with and incorporate critical pedagogy approaches into their course design.

On Practice: Instructional Designer invites us to consider what it means to be an instructional designer, acknowledging all the baggage and buzzwords, but also prompting us to slow down and consider our own humanity in the work we do.

Quality Theater takes a closer look at one of the most prominent online instructional design tools and challenges us to consider more closely how these tools actually impact the success of our students. As an alternative, what if we focus on only those things that matter most: connection, inclusion, clarity, and an overarching theory of practice?

Siberian Syndrome in Online Learning dives into the power dynamics in the classroom and how, when we do not attend to them, students can find themselves on the periphery of the learning community. What happens when we apply this lens to the hybrid, online classroom, where technology creates new kinds of peripheries and reinforces power in new ways?

Building from Our Sensitive Edges focuses on the specific needs of "non-traditional," adult, online learners and proposes an approach that embraces accessible, open, and adaptable practices. These practices allow us to explore mutual accountability, access and inclusion, and collaborative learning communities.

Militaristic Origins, Power, and Faux-Neutrality is an examination of traditional instructional design from the perspective of an education graduate student. In order to imagine a new critical instructional design, we must understand (and grow from) the missteps of the past.

Towards a Critical Instructional Design Framework examines the landscape of critical scholarship related to teaching and learning and synthesizes it into a framework for examining course design. We are asked to consider organization, tools, social relations, and practices but not through a traditional lens; rather, what happens when we interrogate their relationship to learner identity, representation, and power?

Thank you for exploring this volume and giving space for these authors' voices. We hope that your reading brings into clearer focus both the ways in which our current instructional design practices are failing our teachers, students, and institutions and how we can begin to examine, reimagine, and create new practices, approaches, and communities for a critical instructional design.

Counter-Friction to Stop the Machine

Jesse Stommel and Martha Fay Burtis

If you had asked us in early 2020 about the state of instructional design within higher education, we would have told you it was suffering; much of the field, with its roots in military training, had evolved into a landscape of prescriptive models based on a poor understanding of human learning. Private companies were capitalizing on administrative fears in colleges and universities about declining enrollments (Conley, 2019), increased competition (EAB, 2020), "academic integrity" (Turnitin, 2018), and the always-looming existential threat of online education (existential because it rewrote the "rules" of higher education, including the importance of the face-to-face and residential experience) (Quality Matters, 2019). As a result, more and more schools were paying external (usually for-profit) companies to do everything from certifying faculty in a particular ID model (Quality Matters, 2020) to delivering, whole-cloth, courses and degree programs for online or hybrid delivery (Nguyen, 2019).

Meanwhile, at the start of 2020, we would have also bemoaned the state of preparing college faculty for the work of college teaching (Alsop, 2018); in the centuries since the emergence of modern higher education, little has changed to formally prepare faculty during their graduate studies. The state of on-campus faculty development has also been uneven, governed by the willingness of any school to recognize a need for pedagogical expertise among faculty—and to devote resources to the development of that expertise.

Finally, we would have warned about the precarity of our students' lives and just how many are struggling with basic needs, working long hours to pay for their education, and facing enormous debt to pay for their degrees (Rab, 2018). All of this was certainly contributing to more students either not graduating on time—or at all. A concern for students' basic needs is a policy concern but also a pedagogical one. Our material circumstances, students and teachers alike, directly influence the work we do together in classrooms.

As people who have worked in and around the support of higher education for our whole careers, we've been worried for quite some time about these tensions around instructional design, faculty training and support, and student struggles. How do we support our schools in the creation of authentic, well-designed, and human-centered education in the face of such challenges? That was our work. What we couldn't have predicted was the looming threat of COVID-19, or the perfect storm that would emerge as the pandemic swept across our institutions, and we discovered that all these pre-existing challenges were not only heightened by the crisis but, in some cases, our responses to them had been exactly the opposite of what was needed. The crisis has brought into stark focus the many weaknesses in our system: the models we'd depended upon were less resilient than we expected, our staff was spread thinner than we thought, our faculty were less-prepared than we'd hoped, and our students were more vulnerable than we knew. While it might be tempting to dust ourselves off, chalk this up to "experience," and continue on our institutional way, this is a critical moment to take stock and consider what we could have been doing all along that would have helped prepare us better to respond to crisis (this one, the ones we ignored before the pandemic, and the next ones).

The Mire of Instructional Design

The neater and tidier the instructional design solution, the more likely it seems to be broadly adopted by an institution: Learning Styles, Bloom's Taxonomy, ADDIE, Scaffolding, Design Thinking, Quality Matters, Andragogy, HyFlex. Lists, frameworks, Venn diagrams, rubrics, templates. Six principles of Andragogy, five stages of the ADDIE development process, six levels of Bloom's Taxonomy, forty-two review standards of the Quality Matters Rubric. And, even when these models are thoroughly debunked, they continue to retain traction. According to the Association for Psychological Science, "No less than 71 different models of learning styles have been proposed over the years ... But psychological research has not found that people learn differently, at least not in the ways learning-styles proponents claim" (2009, para. 5). And, yet, the American Psychological Association found, "in two online experiments with 668 participants, more than 90 percent of them believed people learn better if they are taught in their predominant learning style" (2019, para. 2) In higher education, too many of us cling to other people's models, because we have rarely been taught, encouraged, or given the support we need to create our own (Stommel, 2020).

Some of the more insidious Instructional Design models are fashioned as needlessly complex in order to create a mystique of intellectual rigor. Even when these models aren't based in research, they are made to seem as though they are. And the worst of these models have strange backgrounds, feed the motivations of for-profit companies, or aim to (or simply do) create edu-celebrities.

Many faculty are first introduced to teaching practices through the lens of Bloom's Taxonomy. There is a certain comfort in the architecture of Bloom's, usually represented as a rainbow-colored pyramid of verbs: remember, understand, apply, analyze, evaluate, create. These are six of the least vibrant verbs we could imagine applying to learning, but this is what learning outcomes so often call for in their pursuit of being *measurable*.

Dull verbs aside, our biggest issue with Bloom's taxonomy is that it's hierarchical. Each level of the pyramid is supposedly built upon the level below. So, you can't "create" or "evaluate" until you first "remember" and "understand." The whole thing feels less like a method to encourage or inspire learning and more a way to police students (and also teachers), laying out a series of hoops for them to jump through with a built-in defense of their existence, what Jeffrey Moro calls "cop shit," (Moro, 2020) or as Jesse has come to call it, "the student agency military industrial complex." In addition, it can feel much easier, especially for inexperienced teachers, to simply map a syllabus to Bloom's than to carefully consider what is fundamentally a messier process.

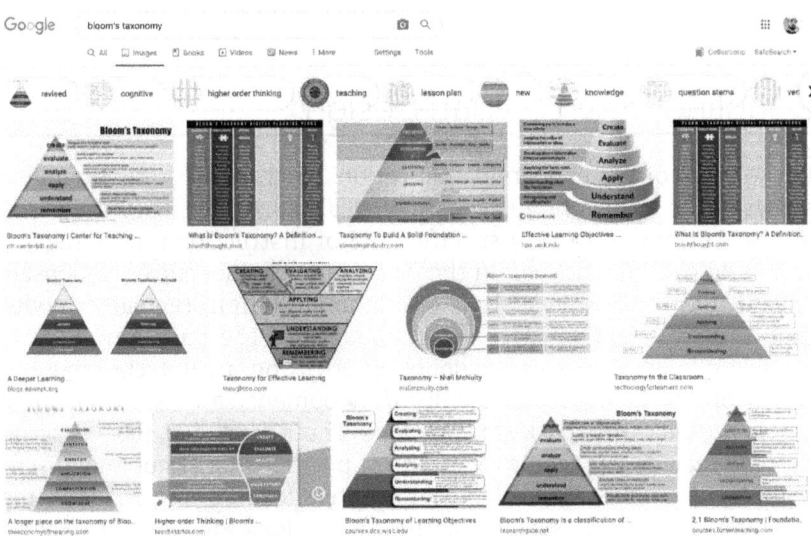

Figure 1: Google Image search result for "blooms taxonomy," showing a range of rainbow-colored diagrams illustrating the theory.

Many of these approaches to instructional design reinforce and rely on systems that oppress students. On the ground at our institutions, the people who are tasked with promoting and supporting these models have little space or authority to push back on oppressive systems. Too often, the schools they work at have consistently devalued their labor and humanity. Instructional designers have long been sidelined (Lederman, 2018). They are alternately given too much power to drive course development, often put into a role of policing the design work of teachers, or they are given far too little power, just as often (and simultaneously) reduced to "service" staff, working precariously and stripped of autonomy. Both structures do incredible damage to the relationship between instructional designers and teaching faculty, pitting them against one another in ways that may benefit an assembly line approach to the production of courses, but only if those courses are stock, standardized, and student-phobic.

Most institutions were either not prepared or prepared in exactly the wrong ways for the crisis of the last year. Stock approaches to online learning can't address the complexities of students and faculty experiencing acute trauma. Standardized course designs don't adapt easily for moments we couldn't possibly have planned for. And fear of students, or a patronizing desire to "help," doesn't create the necessary space for them to find meaning, critically consider their circumstances, or contribute to the building of their education. We aren't aware of a single institution that adequately worked to actively involve students in their COVID response.

Doubling Down on Bad Design

In the wake of the pandemic, institutions are hiring instructional designers at a furious clip (Decherney et al, 2020), but they are hiring almost entirely for a very specific kind of instructional designer, the ones who can help deliver on the promise of cookie-cutter, supposedly evidence-based, models for instructional design, the same models that already failed to meet the challenges of the pandemic "pivot." Institutions are, quite sadly, doubling down on a cafeteria model, in which online courses and faculty development can be served up in small, portable, pre-packaged portions. Three years ago, job ads across the U.S. included dozens and dozens of positions (still not enough) for digital pedagogues, new directors of centers for teaching and learning, and hybrid faculty/ID positions. The options look far more bleak at the moment.

From the job ad for a "Director of Online Learning" at a regional college of art and design: "Reporting directly to the college Provost, this position works closely with our proprietary LMS provider and the college IT department to provide a quality LMS experience for students and faculty." As though teaching and learning with technology can be reduced to a "quality LMS experience." This is one of the grossest phrases we've seen in a job ad. One of the "essential duties" of the job is to "develop and manage the college's online ready-to-run protocols for course checks prior to the start of each term to ensure that course links and videos work, gradebooks are active, assignments are loaded, due dates identified, syllabi and class schedule loaded, etc." This is online learning and course design reduced to its lowest possible common denominator. The word "pedagogy" appears exactly twice in the job ad, in reference to the development of "instructional pedagogy workshops and LMS technical skills training cycles" and "online best practices in pedagogy." There is no recognition or acknowledgment that higher education pedagogy is a field, or that digital pedagogy exists as a robust theoretical and practical set of dialogues and conversations. The idea that pedagogy is purely instrumental, sitting tidily alongside "LMS technical skills," is par for the course.

The complex, brilliant, and diverse human beings with training and preparation in the work of course design, active learning, higher education pedagogy, digital pedagogy, and instructional design are clearly ignored in the creation of a job ad like this one. And these kinds of jobs are increasingly common as institutions scramble to build for and in the wake of the pandemic, while ignoring the collective expertise of their own communities. An instructional design model or quality assurance process is useless (and worse than useless) if an institution fails to listen to and learn from the experiences of its own faculty, staff, and students.

In addition to hiring new positions with uninspired job descriptions, we've also witnessed schools trying to "solve" their current predicament through investment in technology. There is no neat and tidy solution to the challenges we presently face in education. And technological solutionism is the worst possible direction for us to turn. Frankly, it's insulting when institutions throw money at corporate edtech when so many of their most marginalized students are struggling, faculty/staff have been furloughed, public funding has been decimated, and the work of teaching has been made altogether precarious.

If your institution just spent $500,000 on a proctoring solution (a figure recently reported by Drew Harwell in *The Washington Post*), or $200,000, or even $30,000, put that number next to these:

- How many students at your school are food or housing insecure?
- How many faculty or staff have been furloughed, fired, or forced into early retirement?
- How many positions are currently frozen?
- How many faculty or staff have part-time or contingent positions?

Then, add up all the money your institution spends on extraneous (often pedagogically suspect) edtech—like cameras in classrooms, plagiarism detection software, the LMS, proctoring solutions—and compare that number to the total budget for the center for teaching and learning.

Teaching by Accident

Sadly, institutional structures are not designed to cultivate (and sometimes undermine) pedagogical expertise in both instructional designers and faculty. We are excellent at valuing other kinds of expertise, "disciplinary" expertise in particular, but the majority of higher education faculty have little preparation for the work of teaching. After many years completing PhDs in their field, they are expected to enter the classroom and simply embody the role of teacher with little effort or support. It's difficult to pinpoint just how many teaching faculty have had any pedagogical training, because the research on this is quite limited; in 2017, an informal Twitter poll of 2200 respondents by Jennifer Polk, a Canadian academic career coach, unearthed that only 18% of respondents had received pedagogical training in graduate school that they considered at least "decent" (2017). In another informal Twitter poll of 2800 respondents (conducted by Jesse), nearly 60% of higher education faculty said they got a week or less preparation before entering a classroom (2018). A 2013 survey of faculty in the State University System of Florida revealed that only 22% had received any teaching training as graduate students (Hope & Robinson).

How faculty then go about becoming teachers rests on a variety of informal and often unpredictable factors: their own experience as students; how their own teachers taught them; their access and ability to take advantage of faculty development programs at their teaching institution; their availability (and willingness) to engage in a process of examining and iteratively improving their pedagogies.

Meanwhile, at many institutions, when faculty begin teaching online, the landscape shifts. Suddenly, they are required to participate in training to learn the fundamentals of effective online teaching. They may not even be allowed to teach online without getting themselves or their courses certified first. At some schools, they may be assigned a professional instructional designer who will take their existing course content and mold it into a predefined or prescriptive template for online courses. Or they may be required to complete an instructional design training program developed by an external consultant, for which the school pays.

How do we explain this dramatically different approach to preparing faculty for online teaching versus "traditional" teaching? How have we ended up in a place where we expect faculty to osmose good pedagogy and transform themselves into excellent teachers from whole cloth when occupying a physical classroom, yet when faced with a switch to a new modality we scramble towards the other extreme, now requiring experienced faculty to undergo certifications, reviews, and even administrative oversight in their online courses? Why do we so often demean or misrecognize the teaching work done by instructional designers, librarians, advisors, student support staff, etc.?

The answer likely lies in a number of assumptions about teaching that have been baked into our institutions:

1. Good college teaching derives from emulation; faculty can become good teachers because they can (and will) emulate how they were taught. We assume college faculty were taught "well" because they ended up with the terminal degree in their field.

2. Good college teaching derives from good college learning; faculty can become good teachers (again by osmosis) because they understand what it means to be a "good learner." They can translate their own experiences into courses that turn students into "good learners."

3. Since faculty were, by and large, taught mostly in traditional face-to-face contexts, we assume they can only emulate and translate within that modality.

4. Online and hybrid learning are "other" and unfamiliar because they're not how most faculty learned; its pre-

sumed faculty need to learn a new language of teaching or have it translated for them.

At this point, these assumptions have become so institutionalized that most are no longer questioned or critiqued. We have expected faculty to magically transform themselves into good teachers for so long, it does not even occur to us that this is an expectation worth examining. Online learning, on the other hand, is still perceived as new enough (though much less new than most believe) that we continue to hold it at arm's length, treating it as something "other" than our original mission and primary history. Imagine, though, how different your school's response to COVID-19 might have felt if, prior to the crisis, the majority of your faculty had ample support to develop their individual pedagogy (from their first day on campus), had their autonomy and expertise in the classroom been respected and supported, and had abundant and existing meaningful partnerships with instructional designers, librarians, advisors, and other academic-adjacent staff. (We prefer to also call these people teachers.) If this kind of community had existed at your school, would your response have felt less frenetic and would your institution have needed to turn to external solutions?

At the End of All Things

There is a dystopic end point to this fundamental disrespect and lack of imagination, which is already playing out at many institutions. It looks like stock courses, over-architectured right to the edge of oblivion, outsourcing the work of instructional design and online program administration to for-profit corporations, the utter lack of training and preparation in pedagogy for instructional faculty, the increasing precarity of nearly everyone working at institutions of higher education, obligatory use of proprietary systems like learning management systems, increasingly rigid "default configurations" that leave little room for pedagogical creativity, and the proliferation of surveillance tech like plagiarism detection software and remote proctoring. On this last, the CEOs of remote proctoring software have brazenly bragged at the increased adoption of their products, which do fundamental harm to the relationship between students and faculty, as well as the already precarious relationship between faculty and instructional designers.

Over the last 10 months, we've been disturbed to watch many institutions actually cut faculty development budgets and educational technology units while massively increasing spending on LMS contracts, proctoring solutions, plagiarism detection software, cameras in classrooms, and videoconferencing tools. The remote proctoring industry alone "is expected to grow from a $4 billion market in 2019 to a nearly $21 billion market in 2023" (Heilweil, 2020). This is the end point of traditional models of instructional design. This is the corner that crude adherence to rigid course design standards backs us into. This is what happens when we don't support nuanced, complex conversations about pedagogy and design, when we don't valorize the collaborative (and sometimes messy) work of teachers and designers, when we don't design for and with the students who actually show up to our physical and virtual classrooms. In this dystopic future, widespread suspicion of students is the standard, and trust for teachers is eroded. Many institutions are already living in this future.

In "Technology is Not Pedagogy," Sean Michael Morris writes,

> I'm often thought of as the "tech" guy, but what I actually do is very intentionally human. So as I'm approached with questions about what technologies might help build community online, what platform I might recommend for ensuring students don't cheat, or what digital solution I know of that will enable meaningful discussion, I've found myself answering: teach *through* the screen, not *to* the screen. Find out where your students are, and make your classroom there, in a multiplicity of places. (2020, para. 3)

This is the job all of our institutions need to be hiring for. This is the kind of person all of our institutions and administrators need to be seeking out. In far too many cases, they're already a member of our communities, and currently being ignored.

Moving forward, we need a new approach to instructional design, a critical instructional design, not just in terms of what we value (flexible and adaptive, stochastic and dynamic, equitable and just) but also in how we work, the collaboration that faculty and instructional designers should be doing, together. Faculty need to be given the opportunity to learn they are capable of designing across a spectrum of modalities, not just in traditional classrooms and not just for fully online classes. Their need for support, however, doesn't signal that they must also be policed (or erased) from the work of designing and developing courses. Instructional designers are teachers and need to be partners and champions in this work, researching and introducing new models and techniques, co-teaching, peer-reviewing, and engaging in non-hierarchical dialogue throughout. Our teaching mission needs to be understood holistically, playing out across a spectrum of modalities, each of which demand nuanced consideration but none of which are wholly unrelated to any other. Our investment in technology must be critical, careful, cautious; the code inscribed in our tools must be in conversation with our mission. We have to move away from learning objectives, course templates, and technological infrastructures, and instead build community. We need to center people in this work.

References

Alsop, E. (2018, February 11). Who's teaching the teachers? *The Chronicle of Higher Education.* https://www.chronicle.com/article/whos-teaching-the-teachers/

American Psychological Association (2019, May 30). *Belief in learning styles myth may be detrimental* [Press release]. https://www.apa.org/news/press/releases/2019/05/learning-styles-myth

Association for Psychological Science (2009, December 16). *Learning styles debunked: There is no evidence supporting auditory and visual learning, psychologists say* [Press release]. https://www.psychologicalscience.org/news/releases/learning-styles-debunked-there-is-no-evidence-supporting-auditory-and-visual-learning-psychologists-say.html

Conley, Bill (2019, September 6). The great enrollment crash. *The Chronicle of Higher Education.* https://www.chronicle.com/article/the-great-enrollment-crash/

Decherney, P. and Levander, C (2020, April 24). The hottest job in higher education: Instructional designer. *Inside Higer Ed.* https://www.insidehighered.com/digital-learning/blogs/education-time-corona/hottest-job-higher-education-instructional-designer

EAB (2020, February 5). *Colleges signal growing willingness to recruit other schools' students* [Press release]. https://eab.com/insights/press-release/enrollment/colleges-recruit-from-other-schools/

Goldrick-Rab, S. (2018, January 14). It's hard to study if you're hungry. *The New York Times.* https://www.nytimes.com/2018/01/14/opinion/hunger-college-food-insecurity.html

Harwell, D (2020, November 12). Cheating-detection companies made millions during the pandemic. Now students are fighting back. *The Washington Post.* https://www.washingtonpost.com/technology/2020/11/12/test-monitoring-student-revolt/

Heilweild, R. (2020, May 4). Paranoia about cheating is making online education terrible for everyone. *Vox.* https://www.vox.com/recode/2020/5/4/21241062/schools-cheating-proctorio-artificial-intelligence

Hope, W. and Robinson, T. (2013, August 1). Teaching in higher education: Is there a need for training in
 pedagogy in graduate degree programs? Research in Higher Education Journal, 21. https://files.eric.ed.gov/fulltext/EJ1064657.pdf

Lederman, Doug (2018, October 31). Professor, please meet your instructional designer. *Inside Higher Ed.*
 https://www.insidehighered.com/digital-learning/article/2018/10/31/survey-professors-shows-surprising-lack-awareness-instructional

Moro, J. (2020, February 13). Against cop shit. *Jeffrey Moro.* https://jeffreymoro.com/blog/2020-02-13-against-cop-shit/

Morris, S. M. (2020, June 10). Technology is not pedagogy. *Sean Michael Morris.* https://www.seanmichaelmorris.com/technology-is-not-pedagogy/

Nguyen, T. (2019, April 19). Revenue from online ed is on the rise. So is the competition, Moody's says. *The Chronicle of Higher Education.* https://www.chronicle.com/article/revenue-from-online-ed-is-on-the-rise-so-is-the-competition-moodys-says/

Polk, J. [@FromPhDtoLife]. (2017, October 21). *Did you get any teacher / pedagogical training in grad school for TAs or instructors, before you taught or facilitated in classrooms?* [Tweet]. Twitter. https://twitter.com/FromPhDtoLife/status/921790123519434753

Quality Matters (2019, January 30). *Why quality matters in education: The quality difference in 2019 & beyond* [Press release]. https://www.prunderground.com/why-quality-matters-in-education-the-quality-difference-in-2019-beyond/00146341/

Quality Matters (2020, May 6). *University of Southern Indiana Course Becomes 10,000th Course Certified by Quality Matters* [Press release]. https://www.prunderground.com/university-of-southern-indiana-course-becomes-10000th-course-certified-by-quality-matters/00188079/

Stommel, J. (2018, May 15). *Dear higher education teachers, a poll. Answer below, reply with stories, and pass along. How much training in teaching or pedagogy was/is included in your graduate program?* [Tweet]. Twitter. https://twitter.com/Jessifer/status/996355913756893184

—. The human work of higher education pedagogy. *Academe* (Winter 2020). https://www.aaup.org/article/human-work-higher-education-pedagogy#.YtXYzG_MK3W

Turnitin. *International day of action draws attention to the growing threat of contract cheating* [Press release]. https://www.turnitin.com/press/international-al-day-of-action-against-contract-cheating-2018

Centering the Margins Through Critical Instructional Design

Amy Collier, Sarah Lohnes Watulak, and Jeni Henrickson

The COVID-19 pandemic created significant challenges for many, if not most, students in higher education. While many of the challenges students faced existed pre-pandemic, the pandemic illuminated and exacerbated inequities both in our institutions and in historical models of instructional design. Stommel and Burtis (2021) observed a trifecta of educational design issues that had long been simmering pre-pandemic: (1) the historical grounding of instructional design in hierarchical models that promote oppression; (2) the often overlooked basic life needs of students; and (3) a lack of faculty training and support.

The pandemic brought more human-centered instructional design approaches to the forefront. These included trauma-informed teaching strategies, and strategies that centered on care and community, flexibility, and student voice and choice. These approaches form core parts of our own pedagogy, teaching, and writing, and influenced our approach to the study we share in this chapter, which investigated students' lived experiences of the forced transition to remote learning necessitated by the pandemic. In particular, we explore specific structural barriers and challenges students faced during that transition and share ideas for how we might use inclusive design and design justice to bring marginalized students to the core of our instructional designs – not just in reaction to a global crisis, but in an ongoing way.

As members of Middlebury's digital learning organization, we wanted to understand the experiences of students who faced challenges during the pandemic, and particularly during the moment of the rapid transition to remote learning, and how institutional and faculty response shaped those experiences. We recruited participants from two campuses of our institution, an undergraduate college on the East Coast of the United States, and a graduate institute on the West Coast. We identified possible participants through targeted recruitment using data collected by the institution in surveys sent to students in Spring 2020. The surveys asked students to identify needs for support during the transition, including and beyond technology support. We sent an invitation to participate in the study via email to students who reported challenges, and four undergraduate and six graduate students agreed to participate.

For the research design, we chose a post-intentional phenomenological approach, which emphasizes, "the various ways that phenomena are socially produced in context" (Valentine, Kopcha, & Vagle, 2018, p. 466). The approach recognizes and accounts for the specificities of a phenomenon as produced in a particular moment in a particular time and context — in this case, in the context of a global pandemic. Study data were collected via two thirty-minute semi-structured interviews conducted via video conference, a written narrative shared by participants, and researcher reflexivity journals. We took a whole-part-whole approach to data analysis, completing line-by-line readings and coding of the data. Via this process, we crafted tentative manifestations and essence statements (Pazurek-Tork, 2014), and later, a full narrative of the students' shared experience of the transition to remote learning. In Section 2 of this paper, we unpack the main elements of that experience. Frequent research team discussion, as well as individual reflection on our assumptions and positionality vis à vis the study, was also woven throughout the process.

Here, we share the story of one student, Nina, which illustrates the early days of the pandemic shift to remote learning for some students. The narrative is largely in her own words, with some editing for length and to protect privacy. All participant names shared throughout the chapter are pseudonyms.

> Spring break 2020 wasn't much of a break. One day melted into the next, full of anxiety and unhelpful emails, without much sleep in between. I spent the first five days of break driving west from campus to where a friend of mine was living. Going to my parent's house wasn't an option. My plan was to just bounce around from backyard to backyard, campsite to campsite, living out of my car.
>
> As stressed as I was in the days immediately following the announcement that campus was closing as I tried to figure out where to go, arriving at a destination didn't bring much relief. Once I pitched my tent in the dusty backyard of a friend's rental, it was immediately apparent that finding a place to stay was the least of my problems. My friend didn't have internet, and my phone had limited data. I was looking at my bank account, trying to figure out how I was going to afford food, gas for my car, and also buy enough data to connect to online classes.

All the places I might have been able to connect under normal circumstances — coffee shops, libraries — were closed in the initial pandemic shutdown, so I was looking at having to buy a new data plan for my phone, which would double my expenses.

My college had promised that we'd be refunded room and board, but that didn't seem like it'd be back in my pocket for a while. I hadn't budgeted for living expenses this spring besides those associated with living on campus, and I estimated I had less than two weeks before I would have to start exploring the offerings of local grocery store dumpsters.

I spent a day talking with everyone I knew who might be able to help me find some sort of remote work. I had no idea if I'd be able to manage my full course load while working the hours necessary to have any sort of meaningful income. I didn't get the chance to find out because in the uncertainty of lockdown and projected economic downturn, no one was hiring.

The days before classes started were dominated by a frustrating swirl of emails with deans and professors, some of whom were proactive about reaching out. I read but didn't respond to emails from professors who were asking students about our living situations, our availability, our internet access, because I didn't know what each new week would bring, and I was too embarrassed to share my own personal situation over and over again.

Eventually my advisor and another professor close to me were looped into my situation and got me some cash for living expenses from a fund I hadn't known about, and a dean found a bit of money for a mobile hotspot and data plan — not enough to cover the full cost, but with the other cash, it was enough to get to the end of the year.

The rest of the semester was not easy, but the low point was the second week of spring break when I genuinely doubted I would be able to finish my coursework. Not just because I didn't think I had enough data to get to the end of a month of Zoom classes, but also because it seemed like I might have to focus entirely on finding a job and working rather than finishing my studies. I'm so grateful to several of my professors who worked with me to create low-internet versions of their class, who were flexible with deadlines when I was on the road, and who helped me get funding to get through the semester.

Students' lived experience of the transition to remote learning is ultimately a story of connection and disconnection. While technology barriers inhibited students' access to courses, faculty, and friends, and were a source of ongoing frustration, the essence of students' experience of the transition to remote learning isn't "students hated Zoom." Rather, it's a story of trauma, disconnection, and loss.

In the remainder of this chapter, we will explore harms that students faced in their learning experiences, sharing their stories in their own words; and we offer suggestions for how we might counter those harms through specific instructional design strategies grounded in care and community, flexibility, student voice and choice, and design justice.

Student Experiences of the Pandemic Transition

The lived experiences of students we interviewed painted a bleak picture of lives disrupted by the pandemic. Some students talked about "surviving" the transition — not because they are prone to hyperbole but because they struggled with existential challenges. Students who self-identified as coming from marginalized communities (BIPOC, low income, LGBTQ+) expressed a compounding effect — they already felt challenges associated with their experiences of marginalization and those were compounded by experiences caused by the transition. Those students not only had to deal with changes to their learning, they also had to deal with homelessness, abuse, hunger, and a lack of basic access to needed services, support, and infrastructure. Many faced mental and emotional trauma. Kai shared: "the first thing that went through my head was like I'm about to be, you know, flushed into a much less safe environment for my physical and emotional well being, which would just entirely dampen my ability to focus on school work...That thing was me having to move to a completely new area. I don't know how to drive. I don't have a car. There was no public transport. I had no friends. I have an abusive family. And that, I guess, like background struggle was all that was going through my mind when I thought about how I am gonna continue my life as I know it now as a student."

For many, stability was tied up in access to the institution — its supports, schedule, community, services — and when those were suddenly shut down, students suffered their loss. Casey highlighted how the pandemic revealed inequities in students' lives: "[the institution] is really great at...like not everyone is on the same playing field, but at least I had steady internet, at least I had my own room. At least I had a desk, a library, like these very bare essentials, right? And so going back [home], I did have internet but it was very spotty at times...I didn't have my dedicated workspace...there was a lot of distractions... And then also I think there's like a cultural barrier. 'Cause my family is first generation immigrants. And so my mom was like 'why? Why are you doing that? Like come have dinner.' There was a lot of conflict that added to my, it was like a barrier sort of to being able to completely focus."

While some students reported fewer dire struggles, all noted that the transition to remote learning created significant hurdles that they had to navigate, and quickly. As they dealt with those transitions, students also had to deal with the new reality of their educational experience. Technological barriers were widespread, as were other access issues, like having classes with live sessions in the middle of the night for students in distant time zones. Some of the "technical" barriers students identified were created when faculty did not center the perspectives of remote students, which became especially problematic in the fall semester, when some students returned to campus for courses and some remained remote. Jin described challenges with that mixed-modality format: "whenever the professor would turn towards the white board and start writing, his audio, the quality would become worse, simply because he wasn't facing the mic. And you know, the camera's [facing the front] and so when he's engaging with the classroom, I can't see the other students and like that's a little weird...it also just sort of makes you feel a little marginalized, almost as if you're not really getting the entire experience because there's an entire classroom of people actually sitting there."

For many students, the transition forced a reorientation of their time, and for students who were in distant time zones, time was a continual barrier to participation, access to learning resources (such as office hours), and community. Jin, who lived 9 ½ hours ahead of his campus time zone, connected his time-zone challenges to his struggle to manage workloads: "I would write that email in my night, and in the next morning I would receive a response and then I would write an email back to the professor, and then I would receive a response either in night or the next morning...it wasn't sustainable. It was very frustrating, especially because you know, I was spending more hours doing things that could have been easily resolved in office hours."

While some students appreciated how their part of the institution managed the transition, others expressed frustration over how support, communications, and policies were handled. Opportunities for support were difficult to identify, and students like Nina were often "bounced around": "I had not budgeted to be buying my own food until June, right? And so if they could have refunded us right away, that would have been fine, but they couldn't....So then this professor and through this fund I was able to get money to get me through the rest of the year until I could start working...but that wasn't advertised or sent around, [it] took a professor talking to another professor." She added: "I think if [the institution] cared about our mental health, they would have given us an A/A- grading system. I think if they cared about our mental health they would have given students on financial aid 400 bucks when they closed campus...for getting home...for getting through the next few weeks. Like things would have been so much more for my mental health than having someone to talk to. And I think [the institution] thinks that it's just counselors, and it's group therapy sessions, and resources about self-care and puppies at finals. But if they actually, like that, 'cause what stresses us out during the pandemic are having places to live, having food to eat...I think that is the essence of what I wish the school had just focused on is all those material things."

Changes to community may have impacted students the most in terms of their educational experience. Many students reported feelings of isolation and loneliness, and of missing the community of friends and professors they had previously enjoyed. Some struggled with learning in isolation and would sometimes take extraordinary measures to join live class sessions (like waking up at 2am, or driving to a restaurant with free WiFi). Students craved connection with professors and classmates. Some found that connection through live sessions that were structured around dialogue and community building, rather than lecture, office hours, and group work. Graduate students in particular struggled with the amount of group work assigned and with coordinating projects with classmates around the world.

Even when access to peers remained relatively unchanged, for students who stayed on campus or with roommates with whom they were living pre-pandemic, relationships sometimes changed, shaped by the need to now learn, sometimes work, and live together. Val highlighted how isolation compounded her feelings of marginalization: "as a person of color at [the institution], I was more mentally challenged, psychologically too, because it was already challenging as in-person. And then there was a realization that we're going to go online. And that means probably less connection with people. Especially with mental health, like I needed, personally...I would love to have seen my friend in the program and 'cause you know we were always together at the library. We are able to like vent and then work on stuff so that transition of like, 'oh we're going to be online,' it already feels disconnected."

As students navigated new educational experiences, and the disruptions in their lives, many were unmotivated to complete work, participate in projects, and participate in non-curricular community activities offered by the institution. Ana shared: "I realized that I wasn't even caring. I just felt as if, 'oh, fine if I don't submit, I don't submit'... So then [the deadline] would push me to do it and then I would sit through it and finish it on time. But until then my attitude was basically 'I don't care about this. This is not important.'" Lack of motivation may also be tied to mental health issues faced by students. In addition to the trauma of homelessness, abuse, and financial strains reported by some students, most students reported loneliness and disconnection, headaches and other physical problems, worry about their health and the health of family members, the stress of navigating family and friend relationships changed by the pandemic, and stress associated with their courses. As Sidney noted, "I was also trying to think about like, if I do get sick or if someone in my family gets sick, do I want to be here by myself or do I want to be with them or where do I want to be, so...yeah, just like, a lot, like overwhelming amount of information and thoughts to think through in a very short amount of time." Mental health stressors students faced were diverse and compounding, making it difficult for students to navigate their classes and difficult for faculty and staff to provide sufficient support. For example, Jae shared struggles with screen fatigue, especially during long synchronous Zoom class sessions: "You know, I could usually get to about the hour and a half mark, and then I was like...I'm here, but I'm not here. I'm not mentally here. And some professors were not good about breaks... or their breaks would be like, 'oh OK, we're, we're running out of time, OK, take a three minute break.' And I'm like, 'my brain and my body need more than three minutes away from this screen.'"

Mental and emotional trauma was pervasive and damaging to students. While some students took advantage of counseling services, most noted the limitations of those services, and they identified other strategies they needed to cope. Coping strategies included walking with friends or pets, talking on the phone to loved ones, and using strategies such as turning off their cameras during live class sessions–which several students explicitly named as "self-care." Other students, like Ana, struggled with coping: "There were like two, three days after moving, I would just sit. I didn't want to sit alone and think about what happened but every night when I would go to sleep, it would creep in. And then I was like 'Oh no,' I would brush it aside and then start thinking about or start studying for my finals." Students who did take advantage of institutional support for mental and emotional wellbeing felt frustrated by its limits. Trina shared, "I was very sad that we were only given 6 free sessions for [therapy]. So I was always keeping track of which session this was…How can I get the most out of these six sessions…I was also panicking because of the limit."

Students noted, and appreciated, the efforts that many (although, they reported, not all) faculty made to be empathetic and flexible during the transition, and students had empathy for professors who were also having to learn new ways of teaching. Students pointed to the flexibility around assignments and deadlines that professors gave during this time as an example of empathy in action, and appreciated when faculty put effort into organizing Canvas sites and adapting teaching to include synchronous and asynchronous activities. Students also shared frustrations with faculty who were less willing to be flexible, or who provided flexibility on a case-by-case basis: "some of the professors would, if students asked, be a little bit flexible, but it's a lot of work and I have a couple of issues with that. For one of my classes, 10 other students and I had to go on a Zoom to talk to the professor about the ridiculous requirements of the class and how it was unreasonable. And even then we were told like, 'no this has worked for 10 to 15 years, you're not working hard enough' and all this stuff…that doesn't feel good…It was confusing and I think one of my professors would push back deadlines if a lot of students would ask but also that's putting the burden on the students to reach out when it's really hard to do that, and it takes a lot of courage and a large amount of being overwhelmed."

Countering Harms Through Critical Instructional Design

At the moment of the pivot to remote learning in spring 2020, and throughout the academic year 2020-21, many newly adapted remote courses replicated and amplified existing inequities. It's not uncommon for faculty and instructional designers to design for the "mythical average" (Holmes, 2018), treating the experiences of marginalized students as "edge cases" (p. 95-96). Costanza-Chock (2020) cautions that designing for this mythical average, or what they call the "imagined user," reproduces inequality because "designers tend to unconsciously default to imagined users whose experiences are similar to their own... in the United States, this means (cis)male, white, heterosexual, 'able-bodied,' literate, college educated, not a young child and not elderly, with broadband internet access, with a smartphone, and so on." In the case of remote courses, the imagined user or mythical average was assumed to be students who had access to basic needs such as food and a stable living environment, reliable internet access, a computer powerful enough to deal with streaming video and multiple applications, ability to interact with that computer for long periods of time, and the ability to access all needed technologies, among other things.

What would it have been like to instead design with "edge cases" in mind? What strategies can designers of learning experiences use to center the experience of students who are excluded by designs that create barriers to full participation in their learning? In this section, we explore how we might use tenets of critical instructional design, which "prioritizes collaboration, participation, social justice, learner agency, emergence, narrative, and relationships of nurture between students, and between teachers and students" (Morris, 2017), to counter some of the harms that students experienced during remote learning.

We offer four instructional design practices that respond to specific harms articulated by students in our study:

1. Attend to the material and embodied nature of learning
2. Build in flexibility as the default, not the exception
3. Design for low-bandwidth users
4. Combat isolation by creating connection within and beyond the course

For these practices, we include suggested strategies for putting practice into action. Practices and strategies are informed by critical and feminist pedagogy, inclusive design, and design justice frameworks that emphasize centering student voice, and student participation and agency in their learning.

It is important to note that when we design for the "mythical average," students who are excluded must advocate for themselves in order to fully participate in their learning. In our study, Casey highlighted this burden: "Some students may not feel comfortable divulging their home situations and in doing that, it sort of creates this weird dynamic with professors...Students may think that the professor doesn't trust them...some students may have to...share a little bit too much...

more than what they're comfortable sharing with the professor." The strategies that we offer below can be used to proactively build into design practices many of the elements that students had to advocate for themselves during the pandemic.

While we are confident that instructional design approaches that center critical and inclusive practices can counter harms experienced by students, we also feel strongly that some harms students spoke of point to issues that need institutional solutions. Students at the margins of designs need to be centered in design work across the institution. We recognize that instructional designers are typically not positioned within institutions to have the power to shape conversations at that level. We also acknowledge that many of the practices described below have to do with the shape and approach to course content and activities, areas where instructional designers can make recommendations but often do not have the final say in adopting those recommendations. Still, we believe that entering into conversations with faculty about these issues is a beneficial small move toward more inclusive online learning design.

Practice: Attend to the material and embodied nature of learning

In our study, we heard from students who were harmed by policies and designs that did not account for the ways in which material lives and embodied selves intersect with learning. Students' concerns about their material and embodied needs – a safe and quiet space to work, money for food and shelter, access to the internet, breaks from the computer – were not adjacent to students' learning, they were necessary to learning. Posthumanist critiques of dis/embodiment and the lack of attention to materiality in digital learning argue that such things are integral to the co-construction of student learning (e.g, Bayne et al., 2020). Gourlay (2021) unpacks this idea further, laying the groundwork for design approaches that account for the cognitive, social, and material aspects of learning:

> "I would suggest that shifting the focus of practice towards the material and embodied allows us to view digital engagement as fully entwined with it, rather than seeing the material and embodied as context, or means to practice. This allows us to conceptualise digital knowledge practices in a more holistic manner. Issues of access to suitable devices and spaces then become part of practice, and can be brought to the fore.

In pedagogic terms, greater prominence can be placed on the embodied nature of practice, thinking about length of time on screen, people's physical needs for movement, physical comfort and breaks. The priorities of students with disabilities could be made more explicit. The (often gendered) reality of domestic contexts could potentially be made less hidden, and more integrated" (p. 64).

Attending to the material and embodied nature of learning means employing strategies that take into account the embodied experience of learners, strategies that invite students to bring their whole selves to class, and that seek to understand, give voice to need, and provide agency to fully participate in learning.

Strategy: Basic needs support

The Hope Center reported that during the pandemic, "58% [of students] were experiencing basic needs insecurity," with students of color more likely to experience food and housing insecurity (2021, p. 32). However, "the percentage of students experiencing some form of basic needs insecurity was not meaningfully higher in 2020 than in prior years" (2021, p. 36), pointing to the need to provide ongoing support. We suggest including a basic needs statement in the syllabus that provides information about support available at your institution. The Open CoLab at Plymouth State University (2020) reminds us to "be prepared to serve as a resource for students as they navigate challenges related to what you include," and offers an excellent guide to becoming a resource for your students.

Strategy: Make space for students to check in with themselves at the start of class

This strategy was inspired by Andratesha Fritzgerald, author of the book *Anti-Racism and Universal Design for Learning*, who suggests asking your students at the beginning of class "what do you need to be successful or present in class today?" This brief but powerful question invites students to check in with themselves and set themselves up for success based on their needs at the moment. For a synchronous class session, that might mean showing up in pajamas, or keeping the camera off. It might mean making sure to have a snack and a drink handy. The question honors the individual needs of each student based on the ways in which the material and embodied are present and intertwined with their learning experience. In an asynchronous class, you might ask a version of this question at the start of each week.

Strategy: Build in movement and attention breaks

We heard from many students that they worked in front of their computers for 12 hours a day. It's easy to forget that there's a body and brain, which need to move and take a break from time to time, attached to the fingers that are typing on the keyboard. Build in regular movement or attention breaks in synchronous class sessions. In asynchronous course environments, you might create an activity within a module that invites students to take a break, or move around and stretch, etc.

Practice: Build in flexibility as the default

All students in our study noted their need and appreciation for flexibility in courses. Flexibility was inconsistently available to students; some professors provided flexibility upon request, others provided little to none, even when students requested it. Rarely, flexibility was built in from the start and available to all students independent of needs. While students desired flexibility, they also shared how important structure and consistency can be. One student noted that she was more successful in courses with well-defined deadlines because they kept her moving forward. Navigating tensions between structure and flexibility can be challenging, but designing for structured flexibility may provide students options while also offering clear paths to help students manage course workloads.

Why might faculty resist offering flexibility from the start? Some faculty may worry that offering flexibility creates additional work for them, either in the design or assessment of the course. They may worry that providing flexibility will lead to students' taking advantage of them, leading to overly authoritarian and inflexible designs (Denial, 2019). The strategies in this section highlight how flexibility can be part of a "pedagogy of kindness" (Denial, 2019) that supports students in achieving academic goals.

Strategy: Rethink syllabi policies

In the post "Cruelty-Free Syllabi," Cheney (2019) encourages faculty to review their syllabi and ask the following questions:

- What's the tone of [my] syllabus? Do negative commands overwhelm positive invitations?
- Is the premise of the syllabus that students are untrustworthy?
- Are [my] policies designed to punish more than to support?

Denial (2017) shared that her "old syllabus suggested that students had to be told what to do or they'd mess it up. It communicated that they should passively receive my instruction, and it gave them no credit for their intelligence, integrity, or creativity." Cheney (2019) recommends using language that invites students to talk to their professors, language that acknowledges that their lives are filled with other commitments, and language that helps students understand the value of attendance, participation, and on-time assignment submission.

Strategy: Offer a participation menu

Inspired by a faculty colleague and by principles of Universal Design for Learning approaches for multiple means of engagement (CAST, 2018), we have used a participation menu that provides students with multiple ways to participate in class. Participation menus can offer opportunities to interact with others in the class and to do individual work that counts as participation. They might also offer synchronous/scheduled and asynchronous ways to participate, to help accommodate students who want and can connect synchronously and those who cannot. Students also noted that they appreciated flexibility in assignment options, such as being able to submit assignments in different formats (podcast, video, text) and having options for assignment topics or foci.

Strategy: Rethink grading

A rigid approach to grading, often driven by a desire for "rigor," can restrict how much flexibility faculty feel they can offer students. Consider approaches to grading that increase flexibility, such as ungrading (Denial, 2017) or contract grading (Warner, 2016). These approaches provide options for students to decide how much work they want to complete, to reflect on their learning and the quality/quantity of their work, and to engage collaboratively with the professor about their grades. There is a bonus benefit, too, as faculty who have used these approaches say that they have made grading "fun" and that they don't plan to go back to traditional approaches to grading (Denial; Warner).

Practice: Design for low-bandwidth users

We heard from students in the study that access to high-speed internet was sometimes a barrier to participation in class. Even when the majority of students in a class have reliable access to high-speed internet, designing for "edge cases" has benefits for all students (as is frequently the case when we design inclusively). In our work, we often draw on the Universal Design for Learning (UDL) framework to argue for designing course content and activities that are flexible in terms of access, bandwidth, and learner preference. UDL's flexible approach seeks to identify and ameliorate barriers to student participation in the learning environment at the moment of curriculum design, often by providing students with choices in terms of how to engage with content and activities.

Strategy: Provide content in multiple modalities

Providing content in multiple modalities is a small move with a big impact, and allows students to choose which modality best suits them, without having to ask. For example, if you have course videos, some applications make it relatively easy to share or extract audio-only versions to offer alongside the videos. If your video is captioned (highly recommended), you may be able to download the captions and turn them into a text transcript that can be shared. While we acknowledge that captioning takes time and effort, we are heartened by the inclusion of AI-assisted captioning tools in many digital platforms that are lowering the barrier to providing more accessible media. When providing images – including infographics – make sure to include a text description of the image.

Strategy: Offer meaningful text-based participation options

During the pandemic, faculty who required students to attend live class sessions frequently suggested that a recording of the session would be an acceptable alternative for students who were unable to attend. We argue that offering a recording of a live session is not an equitable option, as it does not provide opportunities for students to fully participate in their learning, and may still present bandwidth barriers. We suggest offering meaningful text-based alternatives for students to encounter and discuss content, such as social annotation of readings.

Strategy: Carefully consider the amount of synchronous time needed

Our study highlighted another type of limited bandwidth – time. We heard from students that small group work, especially when it required synchronous interaction outside of scheduled class time, was a challenge for students. Small group work can be designed as asynchronous activities using tools like the Groups area of Canvas, Teams channels, etc. If synchronous small group work is needed, consider building time for small groups to meet during regularly scheduled class time.

Practice: Create connection within and beyond the course

Students shared that the most significant challenges they faced during the pandemic transition related to feelings of isolation and loneliness, exacerbated by losses of their academic community. Whether teaching during a time of crisis or not, faculty should intentionally design courses for community while recognizing that marginalized students may already feel disconnected from or apart from their class and institution. During the pandemic, the notion of "building community" was often associated with synchronous Zoom sessions, which privileged students who had time, access to technology/internet access, time zones close to those of the campus, and an interest in connecting synchronously with others. Even for those students, Zoom fatigue often interfered with effective community-building inside and out of classes.

Strategy: Regularly check in with students

When we asked faculty and students about what worked well during the transition to remote learning, many highlighted how they did more check-ins, such as one-on-one calls or meetings, surveys, and conversations during class. Some faculty mentioned they plan to continue those practices to better understand and support students going forward. Part of antiracist pedagogy's tenet of validation and equity-centered pedagogy, these practices can help marginalized students in particular feel supported and engaged in their learning environment (Ahadi & Guerrero, 2020).

Strategy: Provide multiple ways to connect

As faculty transitioned to remote learning, we heard many lament how challenging it was to create community in online/hybrid courses, given how "natural" and "easy" it was in face-to-face courses. We wonder how many of those "natural" communities are also "imagined" communities, best serving "unmarked" students (Costanza-Chock, 2020). Zoom was the tool of choice for faculty who wanted to replicate in-classroom dynamics; some used Zoom's breakout rooms for small-group discussions. Inclusive design and design justice invite us to ask, "for whom is that approach to connection and community-building best suited, and for whom is it not?" Beyond synchronous interactions, look for asynchronous ways to create community in your class, including ways that are designed for and informed by marginalized students, such as digital storytelling using prompts such as "where are you local?" or "what's the story of your name?" (Bali, 2020; Selasi, 2014).

Strategy: Let students lead community efforts

Building on participatory design approaches, create opportunities for students to lead community-building efforts. This may start with having students to co-construct community guidelines or norms for course interactions. Students can create and facilitate discussions in the LMS or community space (e.g., Slack). Students can bring creative energy and diverse perspectives — building class community through art, music, storytelling, and more. Encourage students to develop multiple means of participation, including analog approaches and embodied approaches (e.g., using their voice, using movement).

Conclusion

All of the students with whom we spoke faced challenges during the pandemic shift to remote learning, ranging from technology to time to workspace to support structures, mental health, and safety. Notably, students who already felt marginalized from the dominant culture and discourse of the institution — particularly students who identified as international or BIPOC — experienced a compounding of marginalization, impeding full participation in their learning experience. Students' basic life needs were sometimes overlooked in the rush to go online, and assumptions were at times made by administrators, faculty, and instructional designers alike, tied to a "mythical user," that burdened students with additional trauma.

The strategies we outlined above serve as a first step to address some of the barriers and traumas faced by the students we interviewed. These barriers existed before the pandemic and were exacerbated by the pandemic. The recommendations in this chapter can and should be extended beyond the pandemic as part of an ongoing pedagogy of care for our students. We can apply these suggestions to all course design projects in the hopes of avoiding future harms.

Significant shifts in instructional design practice such as these will require us to more intensively and intentionally explore what it might look like to build these types of approaches into "standard" instructional design practice as a whole, and to connect instructional design as a profession with inclusive design and design justice. This requires a shift from behaviorist roots and practices (Bradshaw, 2018; Watters, 2021) to a mindset of working openly, co-creating with students, and embodying participatory design, while working to make antiracist and inclusive practices part of the broader curriculum.

References

Ahadi, H.S. & Guerrero, L.A. (2020). Decolonizing your syllabus, an anti-racist guide for your college. *Academic Senate for California Community Colleges.* https://asccc.org/content/decolonizing-your-syllabus-anti-racist-guide-your-college

Bali, M. (2020). *Community-Building Online – Open Resources from @OneHEGlobal & @Unboundeq.* https://unboundeq.creativitycourse.org/activities/community-building-online-open-resources-from-oneheglobal-unboundeq/

Bayne, S., Evans, P., Ewins, R., Knox, J., Lamb, J., Macleod, H., O'Shea, C., Ross, J., Sheail, P., & Sinclair, C. (2020). *The manifesto for teaching online.* The MIT Press.

Bradshaw, A. (2018). Reconsidering the instructional design and technology timeline through a social justice lens. *TechTrends,* 62, 336-344. https://doi.org/10.1007/s11528-018-0269-6

CAST. (2018). *Universal design for learning guidelines* (Version 2.2). http://udlguidelines.cast.org

Cheney, M. (2019). Cruelty-free syllabi. *Finite Eyes.* https://finiteeyes.net/pedagogy/cruelty-free-syllabi/

CoLab at Plymouth State University. (2020). *Basic needs syllabus integration.* https://colab.plymouthcreate.net/ace-practice/basic-needs-and-gateless-policies-syllabus/

Denial, C. (2019). A pedagogy of kindness. *Hybrid Pedagogy*. https://hybridpedagogy.org/pedagogy-of-kindness/

Denial, C. (2019). What do our syllabi really say? *Cate Denial | Blog*. https://catherinedenial.org/blog/uncategorized/what-do-our-syllabi-really-say/

Denial, C. (2017). Making the grade. *Cate Denial |Blog*. https://catherinedenial.org/blog/uncategorized/making-the-grade/

Gourlay, L. (2021). There is no 'virtual learning': The materiality of digital education. *Journal of New Approaches in Educational Research*, 10(1), 57–66. https://doi.org/10.7821/naer.2021.1.649

Morris, S. M. (2017). A call for critical instructional design. *Sean Michael Morris*. https://www.seanmichaelmorris.com/a-call-for-critical-instructional-design/

Pazurek-Tork, A. (2014). *A phenomenological investigation of online learners' lived experiences of engagement* [Doctoral dissertation, University of Minnesota]. University of Minnesota Digital Conservancy. https://hdl.handle.net/11299/168281.

Selasi, T. (2014). Don't ask where I'm from, ask where I'm a local [Video]. *TED Conference*. https://www.ted.com/talks/taiye_selasi_don_t_ask_where_i_m_from_ask_where_i_m_a_local

Stommel, J., and Burtis, M. (2021). Counter-friction to stop the machine: The endgame for instructional design. *Hybrid Pedagogy*. https://hybridpedagogy.org/the-endgame-for-instructional-design/

The Hope Center. (2021). #RealCollege 2021: Basic needs insecurity during the ongoing pandemic. https://hope4college.com/wp-content/uploads/2021/03/RCReport2021.pdf

Warner, J. (2016). I have seen the glories of the grading contract. *Inside Higher Ed*. https://www.insidehighered.com/blogs/just-visiting/i-have-seen-glories-grading-contract

Watters, A. (2021). *Teaching machines: The history of personalized learning*. The MIT Press.

Quiddity of Design in [learning + design]

Nicola Parkin

> The real cycle you're working on is a cycle called yourself. The machine that appears to be 'out there' and the person that appears to be 'in here' are not two separate things. They grow toward Quality or fall away from Quality together.
>
> – Robert Pirsig, *Zen and the Art of Motorcycle Maintenance*

The institutional narratives of alignment, efficiencies, and solutions do not invite us to speak of learning design with any depth or feeling. Focusing on its utility and its techniques at the expense of its mysteries and its passions is a seduction of management designed to elicit our performance. We know that the lived realities of our practices are far richer and more enigmatic than mere method.

I speak as an adult educator working across professional and academic expressions of learning design. I pursue learning design for its ontological depths, for its existential shadows, for its ineffable "somethingness," that which is "more than we can tell," as Michael Polanyi (1966/2009) so perfectly puts it. I have come to appreciate that – ironically! – the fuzzy ambiguity of a learning design spoken like this is actually its strength. Why? Because if we are to resist the hollow ascendency of a learning design tied to only what is measurable, rather than for human being and becoming, we need less defined ways of talking about and engaging with our work. My aim then, as Isabelle Stengers puts it is, "not to say what is, or what ought to be," but to arouse an awareness of a different kind (2005, p. 994).

The puzzle before us is [learning + design]—or, reasonably equivalently, [instruction + design], [education + design]. In each equation, it is the design part that is the stranger. We have grown skilled at asking about the "learning" part as we go about materialising its potentials and promises. However, we rarely ask about the "design" part – even though, as Sean Michael Morris (2018) holds, it is vital that we attend to and recognise design as teaching.

Discovering a world of design within one's existing teaching work can be a revelation, as my colleague Jane found out.

> I think for many academics like myself, what's design, they don't really understand it. But my personal experience of it is profoundly liberating. Profoundly. I've been looking for this ever since I started teaching and didn't realise it.
>
> (J. Haggis, personal communication, May 14, 2015)

Yes, opening to what or where is design in our [learning + design] work gives us new perspectives, approaches, and questions. We open our "design eyes," if you like, onto what we are already doing. When we attend to design as such, rather than just the contents and processes of our designing, we are attending to more than what is immediately before us. We see behind, beyond, and within. We might become more attuned to the contexts and conditions for our learning design work, for instance. In other words, being conscious of "design itself" enables us to work more expansively, critically and deliberately.

The point is not to arrive at an agreed position, but its opposite: to open a conversation. When we ask difficult questions about what is learning, what is design, or what is [learning + design], we are advancing the discourses within which those questions are, or could be, asked (Willis, 2017). As a "field" as such, wherein its practitioners might philosophically self-trouble and contest their work, learning design is weak – so weak that, arguably, it falls short of being a field at all – although this critical reader may be a turning point. Learning design, if it is truly concerned with learning, might begin with enquiring more deeply about itself as a practice. But learning design is a child of higher education, which is poor in a philosophy of itself (Barnett, 2017), its advance into maturity rests with each practitioner, and its conduct as a practice is lodged in our conscience. Surely our work as educators is, on one level, to productively unsettle the ideas that frame our doing and being in the world, in favor of a deeper understanding?

Let's begin it now.

Asking About Design As Such

Australian researchers Sue Bennett, Shirley Agostinho and Lori Lockyer (2017) found that when university teachers talk about their learning design processes, they do not report using formal models of designing. Yet, even if teachers do not call what they are doing designing per se, is it reasonable to say that their teaching decisions are acts of design? Yes, say our trio; they claim that what teachers naturally do counts as a form of designing, since the actions of their work seem to share the characteristics of professional designing, such as working within constraints, reflection, and iteration. Peter Goodyear (2015) is more measured, calling teachers' practices design-like (p. 31), because while they might look like design, these practices lack the professional designer's discipline to curtail impulse with circumspection and to see one's work as occurring within broader frames of reference. Goodyear points out that without this disciplined practice awareness in place, teachers tend to make assumptions and rush to implement solutions.

We could say that what is missing in this teaching-as-design picture is a necessary philosophical dimension to the work. The same applies to professional learning designers, for while we might have design in our job title, we too work from our own personal approaches (Kanuka et al., 2013; Kenny et al., 2005), and are just as likely as the design-naïve individual to ignore the philosophical dimension of our work. For anyone designing within an institutional frame, unless we consciously and critically consider how our designing is conditioned by the contexts of its practice, its figure in our educative work is, at base, functional. Design is then merely a means to an end—a deadline, or a standard—or endlessly looped, simply part of our "frenzied rituals of organisational behavior," as Stephen Marshall so expertly puts it (2018, p. 11). Even if we were disposed to ask about the root assumptions of our work, realistically, we rarely have time.

Enquiring into the nature of our learning design work "as design" does not mean we have to engage with the rules and formalities of design methods or learn its jargon. There is nothing more for us to "do." Instead, we can arouse a different kind of awareness in our work, to our work: we can ask about design's quiddity or "whatness;" we can ask what is undesigned in design, and how design shows itself to be when it is unencumbered by our clever theories and conceptions. After all, these are just designs on design. "Design," as such, is too big, too complex and indistinct to be a singular phenomenon, though it may be a confluence of other phenomena, or found somehow in the boundaries between phenomena (Salustri & Eng, 2007); indeed, by its very nature design resists reduction and is always expanding its meanings (Buchanan, 1992). Designers are not rule makers, they are rule breakers —and they deliberately and constantly challenge and unsettle their field (Buchanan, 2001).

The quiddity of design is behind, beyond, and within our designing rituals. It is that which animates from its source, and its gestures and signatures are ours to discover as our own, being already lively in our educative endeavors, though more than we can say.

The work of asking about design is necessarily philosophical, because design is itself at root philosophic. Design reveals and deals with our ways of being in the world, bringing axiological matters right into our charge, and as such, requires us to be thoughtful. Design, says interaction designers Löwgren and Stolterman (2004), "deals with the profound and existential issues in a very tangible way. As a designer, you have to think about the relation between what can and what ought to be done. Design reveals, in its very practical activities, very philosophical questions concerning how people can and should live their lives" (p. 11).

Applied to learning design, a philosophic appreciation of what is design invites us to become involved with the question about how education ought to be. However, philosophical enquiry in our practice is situated, explorative and generative, rather than teleological. Design philosopher Per Galle (2002) puts it this way: a philosophic stance on design enacts a responsible criticality, which strengthens and renews one's work.

Yes, the philosophic stance is active—it penetrates and unsettles, bringing our work into new motion. In design, philosophy (thoughtfulness), action (decisions) and form (artifacts and activities) come into dynamic tensional accord. Without the philosophical element, we are dealing only with form and action, so the philosophic lends the work of it a satisfying substantiveness. Indeed, without the philosophical element, I argue that design loses its essential quiddity; it becomes simply a way to direct processes towards solutions—and this is the job of management, not design.

Philosophy, when brought into relation with form and action, is not lofty, but highly practical. Indeed, philosophy is fundamental to all our work, for it is philosophies that, "underlie our thinking; our social and personal existence; our innovation; and, ultimately, the solutions and the actions we undertake to address the challenges we face collectively and individually" (Konstantinou & Müller, 2016, p. 3).

By way of example: at a service coordination meeting I attend at the university, we have begun—firstly for fun, and now more earnestly—to begin our meetings with snatches of philosophy. Our very practical agenda has taken on a qualitatively richer feel: the agenda is the same, but we reflect more, we make less assumptions, we seek the underlying principle that binds us, and we ask what is important. Not only have I begun to enjoy these meetings immensely, I have become more personally invested, taking on its agenda as my own.

In the philosophic register, the subject of our attention invites our more genuine involvement. For instance, the notion of "quality" in education is ordinarily front-loaded into our practices with unproblematised hierarchical narratives of excellence – but in the philosophic vein, we can ask not just about the metrics and performances of quality, but about what quality means to us, how it appears to us, and how it moves us when we encounter its unsayable depths in our lived experience. For where is quality but in our own embrace of it?

These intuitions and enquiries already live in our work, even while they might be unspoken. Our questions about value and meaningfulness, whether spoken aloud, muttered to ourselves, or felt as a formless yearning for something "more" speak philosophy into utility, and when summoned, can come to our aid whenever we find ourselves besieged by the banal or the taken-for-granted. It is the asking that is important, not the answers, for by asking about what is good and right and true (however we understand that to be) we transcend our work as an education concerned only with effectiveness and efficiencies: a technology of education, as Gert Biesta (2004) so aptly calls it.

By these simple means – wielding the question, and adopting a deliberately thoughtful stance—we can find that, quite without any intellectual difficulty or theoretical basis at all, we are engaging philosophically in our design work. And whenever we engage philosophically in our design work, we are in the same move engaging with design itself. There: we have exhibited the quiddity of design just by asking what it might be.

But the quiddity of design is not something we gaze on "over there," as if we are not personally involved. When we see our practices more critically, we at the same time glimpse ourselves through our practices (Segal, 2015), so that we find, in practice, we are simultaneously "in" and "out" of what we are doing: we see ourselves seeing. In and out are perspectives on the same figure. Each tells us about the other. For instance, like Ron Barnett and Carolina Guzmán-Valenzuela (2017) did, we can ask about the horizons of teaching to tell us more about teaching's intimate spaces. We can also take the opposite stance, and ask interiorly about the outward conditions of our work. We work concomitantly from the inside out, and the outside-in.

Contradiction, movement, and play prevent us from getting caught up in philosophy's straightjackets, and are all ways of taking care in philosophic pursuits. What I mean is – while the philosophic stance on and in one's work might yield strength and depth and perhaps valuable personal and practical insights, it does so ironically, by withdrawing from making any claims. The philosophic, if it is to resist becoming itself a technology, a means to an end, must keep asking about itself—it must keep breaking its bounds and stay endlessly open. In other words, a philosophic approach that becomes too sure of itself, or too automatic, extinguishes itself. (It must, to be fair, not say "must" quite as much as I am doing!) As Peter Sloterdijk (2006) points out, if we are to oppose something, our response must be different to that which we oppose. If we want to oppose a technology of education, we do so by holding openness as sacred in our practices, and favoring humility and uncertainty over assurances and finishedness. In a milieu of arrogant self-assertions, inviting the gentle playfulness and earnestness of a philosophic approach to take root in our practical work is perhaps the most unquieting approach of all.

Design, like philosophy, carries its own traps. Fixate too much on the design part of [learning + design] and there is a danger that we will simply introduce another layer of technicism to learning design work. Quiddity is our friend here, reminding us to attend to design's inner laws, rather than its rules, if you like. We attend to what is behind, beyond and within; what is at heart, what is not immediately apparent, what seems absent or distorted.

On the face of it, one need not look too far for design: design is everywhere, and we are in it. But design does not always declare itself. In fact, design is the master of hiddenness, covering or camouflaging our realities; hiding even itself from detection. Being in design is like being in a giant blind spot. The designed world we are in is largely undisclosed (Bell, 2017); below our line of sight are the thickets of invisible interconnected data, the territories of commercial interests and the wild algae blooms of algorithms that stealthily condition the activities of our work, our learning, and our leisure. Hidden as such within our artifacts and behaviors, it is easy for design to be a work of deliberate deception, hoodwinkery (Flusser, 1999) and craftiness (Singleton, 2014). With design, says Tony Fry, "what you see is not what you get" (2014, p. 12).

Our own designing activities occur largely within conditions and frames of reference that are themselves designed. As teachers, we are granted a seeming authority to design the educational experience as we see fit, but in reality, we are making our design choices within a prefigured and delimited "choice architecture" (Thaler & Sunstein, 2008). The learning management system (LMS) is a prime example of choice architecture whereby our learning design choices occur within a deliberately delimited set of socio-technical constraints configured on many fronts: by the platform "product" itself; by the institutional policies that direct its parameters and use; by the conditions for our choosing, which we may or may not notice and against which we may or may not rail; and more subtly, by the normative practices to which we all, to some degree, conform. The LMS, as William Beasley (2012) says, is a kind of walled garden designed to shut out the outside world. It creates a world within the world.

We can think of layers of design, nested one within the other – design is never alone, but always also behind, beyond and within design: [[design [design] design]]. Indeed, one way of understanding the professional learning designer's operating zone is that of working up and down these design layers, creatively interpreting, reinforcing, transgressing, and developing them in local contexts as contextual needs and personal proclivities dictate. Design is like that: restless. We work from the outside in, and the inside out. We work largely in the dark, but towards the light. Design is at home with legitimate constraints, this is not the issue; what is important is that the constraints are explicit – that way, we can choose how we respond to them. For my own part, responding to the constraints is somewhat of a game that I like to play – I enjoy pushing the LMS as far as I can, using it as a canvas for breaking the rules. It is not the rules themselves, but the hoodwinkery that I object to.

Design's enframing effect works both on and within our own design efforts. We regulate and constrain our own designing with our habits of thinking and doing. For instance, we assume that we ought to finish an idea we are working on, even if it comes at the expense of our contested, emergent, messy realities (the same realities that fertilise the pedagogical). Roll out neat packages of learning, and we leave little to the learner's imagination. There is a case for less design. We have already been told not to over-stuff the curriculum—should we not also pare back the work of trying to manage and direct the experience of learning (as if it were ours to control)? Yes, "too much" design, no matter how intentionally sound, and we risk overwhelming the possibility of an original and spontaneous learning—thus undoing our innermost designs with our utmost designs, so to speak.

A [learning + design] sensibility, where we are mindful of the interactions between [learning] and [design], is surely for the myriad relations between learning and design, rather than a one-way flow. Design does not bring about learning; there is a reciprocity between design and learning. Just as design sways the possibilities of learning, learning sways the possibilities of our designing: it goes both ways. And the two are natural allies, for both learning and design have their ultimate being in what cannot be known in advance. [Learning + design] is not an equation looking for a "=." This pairing must stay tensionally entangled for learning design to be wholly itself: [learning design].

The quiddity of design in our educational attentions is, then, something to do with remembering to notice, ask about, and excite the tensions between design and learning for the sake of learning. If our designing works against learning, even a little bit, it must be abandoned, or at least, dismantled. Design's quiddity is lost without learning's quiddity in its sights. As Anne-Marie Willis (2018) contends, if we are to avoid our design "doing" becoming our "undoing," then our designing necessarily involves counter-designing. As educators, we should take this responsibility seriously.

Design's Summonses

In our design work we come face to face with design's "most uncertain, contradictory, dangerous, and promising summons" (Nelson & Stolterman, 2012, p. 188). This summons comes from without, in the form of great challenges, and from within, as a kind of working conscience. From the outside in, and from the inside out: to design is to let oneself be addressed, and to enter into a kind of back-and-forth dialogue with the world. The question is not an arrow but a space we enter into: when we ask about design, we invite design to ask of us: we invite the summons – and in doing so, declare ourselves already summoned.

Understood this way, design is a disclosive experience that one is "in:"

> The process of design is thus a disclosure, in two senses. Firstly, it is a disclosing of the artefact that is being designed; and secondly, and simultaneously, it is an unfolding of self-understanding, since it reveals one's preunderstandings. It uncovers the preconceptions that are constitutive of the design outcome,

and at the same time brings to light the prejudices that are constitutive of what we are. The design process is an edification in two senses: it builds up the artefact and edifies the designer (Snodgrass & Coyne, 1996, p. 25).

Then – one comes to know design by paying attention to oneself in design? Is this what "being" a designer means? The design scholars point in this direction; they say that a designer is one who intentionally and consciously attends to their practice (Salustri & Eng, 2007), one who responsibly accepts the call to design (d'Anjou, 2011), who is themselves shaped by their designing (Wendt, 2018), and who is aware of the consequential involvements of their designs in the designed world (Nelson & Stolterman, 2012). A designer is one who also knows when a situation is not about design (Mitcham, 2012). None of these understandings are to do with design's rules or conventions, but to having an awareness of oneself in dynamic relation design's own "self," or quiddity.

Conscious of ourselves in design and even "as" design, we lose our design naivety, but preserve the possibility of a genuine "undesigned design" rooted in its deeper layers. As such, we can no longer rely on what can be measured externally, but are thrown on our human mettle: we must rely on the quality of our awareness in practice.

Learning from Learning

The first summons is to learning itself, for in [learning + design], design is only extant by its necessary and abiding entanglement with learning. Since we as educators are already in the place of learning, we can approach design from here. The approach, however, is not a route to a destination, but a stance made within its already-hereness. We begin where we are: involved with learning. And, since learning is never what we expect it to be (Ramsden, 2003, p. 8), and in any case there is no agreement about what learning is (Barnett, 2011), thankfully, there is still room for learning about learning. Learning's appearance cannot be known in advance, and we should not expect our designing of learning to foretell it or compel it—this would be a kind of educational arrogance. We would do well to refrain from even trying to reach an agreement about learning, for not knowing about learning keeps our learning design innocent.

Learning and design exchange forces. In the equation of [learning + design], the "+" stands in for something generative to be discovered, rather than an adding up—we have had enough of a calculative kind of education, after all. The tension between them, rather than being for resolution, brings them both into greater relief. Perhaps [learning "vs" design] might communicate my intention better, although the "vs" does not carry into our imaginations the pedagogical potency in this difficult pairing of forces. It might help if I use Maxine Greene's powerful words from 1992 on the "educational equation" here to press home my own [learning + design] point:

> We are charged, we who care about thinking and teaching, to study that equation and keep trying to discover what does not "add up." We may be able to find connections that enable us to do something about the desire to submerge in a comfortable life, the tendency to believe blindly, the dedication to profit, even the self-infatuation of the few. … It will take critical consciousness, imagination, thoughtfulness of many kinds. It will take the opening of spaces where people can come together, where they can choose. It will take disclosures and refusals and the shaping of new visions. It will take thinking what we are doing, knowing there is no stopping place and that the search must continue on (p. 15).

Aha! Then our equation [learning + design] contains within itself an exquisite irony – for both learning and design love whatever does not "add up!"

We can find new perspectives if we play with this equation. For instance, instead of saying we design for learning, it might be better to say we design from learning. If we are not learning from learning, I think it is because we have, in part, objectified our material by speaking of teaching as the flip-side of learning. This is merely another assumption that we have made; I know this because I was surprised some years ago by a free-thinking colleague of mine who shared her own assumption that "teaching and learning" referred to the integration of teachers' own learning into their practices. Yes – why not think of it this way? It is critical that we inhabit learning from inside the experience of learning, if we are to be experts with our material. Learning must be gained from learning itself: from its own origin, which springs in each of us, teacher and student alike.

Arguably, then, the educator's work is to know that origin in themselves. We must avoid the trap of thinking that learning design is "over there." We are each part of the equation; we ourselves are the brackets […] that contain [learning + design]. We open and close our work at will, ontologically, within our being. Accordingly, we let ourselves learn about [learning + design] from within its lively equation. Even at the level of a field of practice, learning design is still learning about itself, and the field is nowhere, if not lodged in our common attention to its possibilities for our work.

Design Turns Towards Difficulty

If you are finding the quiddity of design a difficult idea, and yet have read this far, then you have already entered into its spirit, for design holds within its ambit a willing embrace of difficulty; indeed, design turns towards difficulty. As such, design is a place where an educator can be at home with personal risk and the embrace of discomfort (Fellmayer, 2018). Instead of thinking about design as a means for closing down risk, design can help us find a way in.

Personally, this appreciation has helped me enormously. When I remember to orient this way, I can understand the tensions that besiege my work "as" my work, rather than something that gets in the way of it. Indeed, when difficulty presents, I am grateful for the crack, for it delivers me from a state of blithe unconcern—after all, it takes a rupture for our practices to explicitly become a theme of concern for us (Segal, 1999). When a schism appears—say between what we personally hold to be important and the institutional narratives of worth that press upon us—then, we pay attention with our very selves, and so embody the tensional equation.

For me, the appearance of difficulty is welcome, for it signals that I am emerging into a deeper engagement or understanding with my work. Sometimes I seek it out deliberately. Difficulty is then transformed from what thwarts my work, to what gives my work back to me. After all, notes philosopher Karl Jaspers, it is the "lively disparities" between plan and outcome that save our work from being purely mechanical (1932/1970, p. 133). Might we then not court some of this liveliness under the banner of our learning design work, and make some good trouble for ourselves? We may be tired and maybe even cynical, but it is worth following the ruptures, opening inward, digging down.

Of course, we might feel drawn to seek out a few securities, and realistically, we are not always practically or professionally disposed towards difficulty. Exhausted by the struggle, I might prefer to concede it, as a form of self-preservation. But if I sidestep the difficult altogether, my learning design work will be dead for me, reduced to box-ticking and rubber-stamping (Sloan & Bowe, 2015); an exercise in conformity only.

Then, I will find myself in a design situation in which what is truly configured is not the possibility of learning, but my compliance.

The point is that we have a choice: perhaps even that the quiddity of design is choice, insomuch as choice contains within itself those elements of philosophy, action, and form. Design, as choice, is strange—it opens and closes in one move.

Design as Our Figure of Freedom

Tony Fry's description of design as our "figure of freedom" (2005, p. 139) seems to perfectly capture and communicate design's powerful opening-and-closing forces, its inherent philosophical necessity, its action-orientation, and its concern with form. Conceiving design as a figure of freedom, our orientation is to work with the tensional forces, both freely creating within limits and curtailing freedom through purposeful delimitation. We also figure new limits by which our designing might perpetuate itself—we figure design's own freedom to be.

Let me break this down through our metaphor of the equation. We can think of what is within the brackets [...] as the "design space;" that is, the field of our attentions and attunements (Akama, 2012). Within this field, we are free. Attending to the brackets themselves as the limits of our own awareness, the conditions in which our designing occurs, and the delimitations that we deliberately impose, our designing contains itself, in us. The constraints and the freedoms of our work are mutual, because the constraints can themselves open the space within which form is conceivable—forms that, once opened, must then be contained – and so on, in the mutual disclosivity (if that can be a word—why not?) of self-and-form.

Yet, at the same time as we (con)figure our freedom within the [design space], "what" is designed always comes out of and exceeds its bounds ...[]..., because our designs open new forms of experience-spaces with their own [...]. Our nested design worlds open as well as contain being's experience spaces in their cascading motion. In designing, I open the world as it never was and bring it into new motion; a world which McCarthy and Wright (2004) explain in Technology as Experience, "though already half-designed, is always becoming;"

> In such a world, design is always for potential, for what is already becoming. It is an act of reframing experience in a way that points beyond the reframing. This involves the designer

giving to the user a surplus, which allows them to play into their potential (p. 196).

The "user" in our design context, is the student. Design is for the world, not for ourselves. It is as if design spins itself out of one's being and becomes part of the fabric of our shared being in the world. In offering design into experience, my work is deeply personal, but at the same time, vast; it exceeds me, and goes wide and long. In this way, design gains a life of its own in ways that do not belong to me: I must give it away for it to be free to do its work (but in giving it away, it returns to me in kind).

Perhaps then, as the figure of freedom, the only thing that I might sincerely "design" as such is that which preserves my designing as a figure of freedom. A dangerous summons indeed!

Design as Disclosing Design

The quiddity of designing is ours to enjoy within the context of our teaching, but its fruits are for the student. It is not a stretch to appreciate that by disclosing design in our teaching, we pass on its figure of freedom as part of pedagogical intentions. Yes, rather than enclose the student in our designs, we pass the design to them; we pass the [...] to the student. Way more than the "surplus" that McCarthy and Wright signal, the uniqueness of learning design as a form of design is that it can give itself away entirely through the pedagogical act of disclosure. For instance, we can offer a reading list, but we can also say why we chose that list: what lenses we used when selecting, and what our intentions are for the student when they engage with those readings.

As educators, we are free to disclose design in the events and affairs of learning. I would go so far as to say that we are obliged to disclose it. We can also choose to reveal the schism, the struggle, the difficulties of our designing in our designs, as our designs. In other words, we can preserve the very human acts of design in the artifacts of design. In this simple but radical move, our learning design is turned inside-out, and "undesigns" its own camouflage. Now, because our own design eyes are open, we can invite students to notice the designs that they are already in; designs that configure their attitudes and choices, and indeed their learning – we can help them open their own design eyes.

We are used to hiding our teaching behind design. Especially in the online learning context, we have come to rely on design to stand in our place when we cannot be bodily present in the event of teaching and learning. We want to prepare our lessons well, hoping not to disappoint, or to be caught out as a fraud. But if we are brave, we can let conversations about the best tools and methods for learning occur as part of the educative experience, rather than thinking we need to prepare these [...] experience-spaces in advance. If we do prepare in advance, we can disclose the design within our work, by explicating the philosophy of our choice, and how it came into action and form. Why keep the way of our work hidden? Why go to all the trouble of designing a lesson that serves only to fulfill only its own pre-configured "learning outcomes?" The only learning we can be sure of here is that we as teachers have learned the rules of properly structuring a lesson. If education's summons is for the appearance of freedom, as Greene (1988) says, then surely we must ourselves exemplify that freedom in our educative practices?

Holding open our designing in our teaching is like teaching naked, without guile or guise. It is intimate and bold, but when we open ourselves sincerely, we also serve learning most sincerely. Disclosing design in our teaching gives power to humility. We cannot be more personally generous than this, and we cannot be more pedagogically generous than this. We can stand together in learning with the students, in a "suspended" educational state, where, "a new kind of freedom can be sensed, a new kind of education explored, a new common manifest, which belongs to no one and resists all forms of regulation and assessment beyond its own appearance" (Lewis & Friedrich, 2016, p. 248).

A transcendent education like this cannot be "designed" as such. Yet, if we do not ask about design, we remain ourselves the subjects of design, even perhaps its unwitting minions, rather than its fiduciary agents —or counter-agents, as needs be. But we are not without direction. It is my conviction that by tapping into design's quiddities, we can open our [learning + design] anew, inside our original experience of it, and so discover something we can believe in to offer into education.

With, of, as, and to.

> *Where it comes from, so do you.*

Note: This chapter draws on, remixes and extends snatches of text from my doctoral thesis: Parkin, N. (2021). *Intimacies of being in learning design* [Doctoral dissertation, Flinders University]. https://theses.flinders.edu.au/view/d85f1904-3007-429d-9a71-e8868960d89d/1

References

Barnett, R. (2011). Learning about learning: A conundrum and a possible resolution. *London Review of Education*, 9(1), 5-13. https://doi.org/10.1080/14748460.2011.550430

Barnett, R. (2017). Constructing the university: Towards a social philosophy of higher education. *Educational Philosophy and Theory*, 49(1), 1-11. http://dx.doi.org/10.1080/00131857.2016.1183472

Barnett, R., & Guzmán-Valenzuela, C. (2017). Sighting horizons of teaching in higher education. *Higher Education*, 73, 113-126. https://doi.org/10.1007/s10734-016-0003-2

Beasly, W. (2012, May 2). Infiltrating the walled garden. *Hybrid Pedagogy*. https://hybridpedagogy.org/infiltrating-the-walled-garden/

Bell, G. (2017, October 21). Boyer lecture: Fast, smart and connected: What is it to be human, and Australian, in a digital world? [Radio broadcast]. *Australian Broadcast Corporation*. https://www.abc.net.au/radionational/programs/boyerlectures/genevieve-bell-fast-smart-connected-how-build-digital-future/9062060

Bennett, S., Agostinho, S., & Lockyer, L. (2017). The process of designing for learning: Understanding university teachers' design work. *Educational Technology Research and Development*, 65, 125-145. https://doi.org/10.1007/s11423-016-9469-y

Biesta, G. (2004). Against learning. Reclaiming a language for education in an age of learning. *Nordic Studies in Education*, 23, 70-82. https://orbilu.uni.lu/handle/10993/7178

Buchanan, R. (1992). Wicked problems in design thinking. *Design Issues*, 8(2), 5-21. http://www.jstor.org/stable/1511637

Buchanan, R. (2001a). Design research and the new learning. *Design Issues*, 17(4), 3-23. http://www.jstor.org/stable/1511916

d'Anjou, P. (2011). An alternative model for ethical decision-making in design: A Sartrean approach. *Design Studies*, 32(1), 45-59. https://doi.org/10.1016/j.destud.2010.06.003

Fellmayer, J. (2018, October 11). Disruptive pedagogy and the practice of freedom. *Hybrid Pedagogy*. https://hybridpedagogy.org/disruptive-pedagogy-and-the-practice-of-freedom/

Flusser, V. (1999). *Shape of things: A philosophy of design*. Reaktion Books.

Fry, T. (2005). On Design Intelligence. *Design Philosophy Papers*, 3(2), 131–143. https://doi.org/10.2752/144871305X13966254124518

Galle, P. (2002). Philosophy of design: An editorial introduction. *Design Studies*, 23(3), 211–218. https://doi.org/10.1016/S0142-694X(01)00034-5

Goodyear, P. (2015). Teaching as design. In P. Kandlbinder (Ed.), *Higher Education Research and Development Society of Australasia (HERDSA) Review of higher education*, 2, 27–50. https://www.herdsa.org.au/herdsa-review-higher-education-vol-2/27-50

Greene, M. (1988). *The dialectic of freedom*. Teachers College Press.

Greene, M. (1992). From thoughtfulness to critique: The teaching connection. In W. Oxman, M. Weinstein & N. M. Michelli (Eds.), *Critical thinking: Implications for teaching and teachers* (Proceedings of the 1991 conference, New Jersey). Institute for Critical Thinking, Montclair College. https://files.eric.ed.gov/fulltext/ED352358.pdf

Jaspers, K. (1970). *Philosophy* (E. B. Ashton, Trans.; Vol. 1). University of Chicago Press. (Original work published 1932)

Kanuka, H., Smith, E. E., & Kelland, J. H. (2013). An inquiry into educational technologists' conceptions of their philosophies of teaching and technology. *Canadian Journal of Learning and Technology*, 39(2). https://doi.org/10.21432/T2KS3B

Kenny, R., Zhang, Z., Schwier, R., & Campbell, K. (2005). A review of what instructional designers do: Questions answered and questions not asked. *Canadian Journal of Learning and Technology*, 31(1). https://doi.org/10.21432/T2JW2P

Konstantinou, E., & Müller, R. (2016). Guest editorial: The role of philosophy in project management. *Project Management Journal*, 47(3), 3–11. https://doi.org/10.1177%2F875697281604700301

Lewis, T. E., & Friedrich, D. (2016). Educational states of suspension. *Educational Philosophy and Theory*, 48(3), 237–250. https://doi.org/10.1080/00131857.2015.1004153

Löwgren, J., & Stolterman, E. (2004). *Thoughtful interaction design: A design perspective on information technology*. MIT Press.

Marshall, S. J. (2018). *Shaping the university of the future: Using technology to catalyse change in university learning and teaching. Springer.

McCarthy, J., & Wright, P. (2004). *Technology as experience*. MIT Press.

Mitcham, C. (2001). Dasein versus design: The problematics of turning making into thinking. *International Journal of Technology and Design Education*, 11, 27-36. https://doi.org/10.1023/A:1011282121513

Morris, S.-M. (2018, April 12). Instructional designers are teachers. *Hybrid Pedagogy*. https://hybridpedagogy.org/instructional-designers-are-teachers/

Nelson, H. G., & Stolterman, E. (2012). The design way: Intentional change in an unpredictable world (2nd ed.). *MIT Press*. https://mitpress.mit.edu/books/design-way-second-edition

Polanyi, M. (2009). *The tacit dimension*. University of Chicago Press. (Original work published 1966)

Ramsden, P. (2003). *Learning to teach in higher education*. Routledge.

Salustri, F. A., & Eng, N. L. (2007). Design as…: Thinking of what design might be. *Journal of Design Principles and Practices: An International Journal – Annual Review*, 1(1), 19-28. https://doi.org/10.18848/1833-1874/CGP/v01i01/37591

Segal, S. (1999). The existential conditions of explicitness: An Heideggerian perspective. *Studies in Continuing Education*, 21(1), 73-89. https://doi.org/10.1080/0158037990210105

Segal, S. (2015). *Management practice and creative destruction: Existential skills for inquiring managers, researchers and educators*. Routledge.

Singleton, B. (2014). *On craft and being crafty* [Doctoral dissertation, Northumbria University]. Northumbria research link. http://nrl.northumbria.ac.uk/21414/

Sloan, A., & Bowe, B. (2015). Experiences of computer science curriculum design: A phenomenological study. *Interchange*, 46(2), 121-142. https://doi.org/10.1007/s10780-015-9231-0

Sloterdijk, P. (2006). Mobilization of the planet from the spirit of self-intensification. *TDR/The Drama Review*, 50(4), 36-43. https://doi.org/10.1162/dram.2006.50.4.36

Stengers, I. (2005). The cosmopolitical proposal. In B. Latour & P. Weibel (Eds.), *Making things public: Atmospheres of democracy* (pp. 994-1003). MIT Press.

Thaler, R. H., & Sunstein, C. R. (2008). *Nudge: Improving decisions about health, wealth, and happiness*. Yale University Press.

Wendt, T. (2018). Arational design. In P. E. Vermaas & S. Vial (Eds.), *Advancements in the philosophy of design* (pp. 101-120). Springer.

Willis, A.-M. (2017). Editorial. *Design Philosophy Papers*, 15(2), 95-97. https://doi.org/10.1080/14487136.2017.1390193

Willis, A.-M. (2018, 17-18 May). *Ontological Design, Criticality and What Comes After Design?* [Conference presentation]. Critical by Design?, Basel.

Compassionate Learning Design as a Critical Approach to Instructional Design

Daniela Gachago, Maha Bali, and Nicola Pallitt

COVID-19 has forced a global emergency pivot to online learning, which has led to increased demand on both educators and learners, with a major impact on workloads, research careers, and mental health. Across the globe, academics complain of burnout, exhaustion, and lack of self-care. This has heightened the concern around the well-being of learners and staff (Czerniewicz et al, 2020; Imad, 2021) and has led to an increase in interest in approaches to teaching and learning that recognise the importance of care and compassion, such as humanising pedagogies (Pacansky-Brock, 2020); pedagogies of care (Bali, 2020) or trauma-informed approaches to pedagogy (Imad, 2021; SAMHSA, 2014; Costa, 2020).

Educators drawing on these approaches are concerned with their learners and their socio-emotional wellbeing. They see learning as happening when we learn in communities and when we feel we belong. Here the learning environment and the way we facilitate learning and community become as important as the course content. As Borkoski (2019) argues, "There is consensus in the literature about the benefits of a student's sense of belonging. Researchers suggest that higher levels of belonging lead to increases in GPA, academic achievement, and motivation"

Designers often hear about "empathy", "care", "compassion", "inclusive design", and "trauma-informed" as part of their professional work. However, there is also a lack of consensus around the meanings thereof, or how accounting for them influences their learning design processes and outcomes. This contribution offers a critical engagement with these concepts, examining the relationships between these and their use in design models and approaches. We provide an alternative model that positions empathy and compassion along a continuum with design considerations that recognise and work with the power relationships between educators and learners to create opportunities for learner agency, participation, and empowerment. Our model draws from affect theory and is informed by perspectives on empathy, care, and compassion but is also underpinned by a social justice agenda.

It involves choices and considerations that instructional designers or learning designers and educators make in their relationships with learners following a continuum of design approaches that exemplify the level of learner participation by designing to, for, with, by, and as learners (adapted from Wehipeihana, 2013).

In previous work (see for example Gachago et al, 2021), we have expanded the traditional instructional/learning design approach by integrating principles and activities from the design thinking literature to develop a learning design model that would put empathy at the centre of all design decisions.

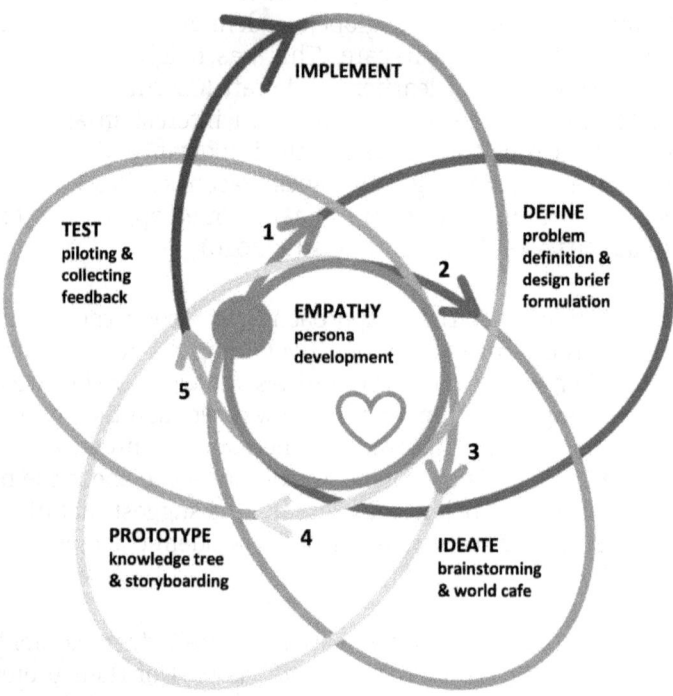

Figure 1: CPUT learning design model (Gachago et al., 2021)

Design thinking traditionally starts with developing a "persona", which is a user archetype, visually represented, that helps to focus a design activity on a particular individual or user, in a specific context (Anvari & Richards, 2018; Van Zyl & De la Harpe, 2014). In this activity, participants are asked to imagine one typical learner in their class (this can be a real learner or a fictitious learner made up of many of the learners they have encountered in their teaching practice) and to describe this learner in detail (including demographic details, educational history, professional career, motivation, strengths, challenges and the like). It is important to give this person a name and to make them as authentic as possible. We use flip chart paper and whiteboard markers and participants present this persona to the rest of the group. Sometimes lecturers express their discomfort about "stereotyping" or "labelling" learners. We take care to address these concerns in a sensitive and empathetic manner (see Van Zyl & De la Harpe, 2014). The persona flip charts are pinned to the wall, serving as a constant reminder of the people we are designing with and for throughout the different stages of the learning design process. These personas can represent current learners and their challenges but also future learners, or graduate personas representing learners who have completed their course/qualification. This is a useful activity to unpack how graduate attributes can be integrated across curricula. However, we have also felt increasingly frustrated by two factors that seem to be implicit in learning design models drawing from design thinking.

1. The assumptions about who learners are (as exemplified by the traditional persona activity, that many of our design processes start with), and designing a priori based on that, rather than in response to who the *actual* learners turn out to be, and how to create designs responsive to those different learners (and learners that might change with each iteration of a course)

2. The use of empathy interviews and personas also does not question positionality or biases of the designers themselves when they conduct and interpret interviews and create and design for personas. Again, the choice of whom to interview, how to interpret this data, and how this impacts the development of personas is not impartial and is not the same as designing "with" the actual learners.

In this contribution, we are exploring what it would mean to replace empathy with compassion, which we see as more concerned with equity and justice, following educators writing about affect in the classroom such as Megan Boler (1999), and Michalinos Zembylas (2008; 2011). These authors argue that feeling empathy for the other is not enough, as it creates both distance (Boler, 1999) and a superficial understanding of difference (Zembylas, 2008). Rather, these authors call for a more critical engagement across difference, one that recognises the need to engage with difference both on a personal but also systemic level, emphasising a collective responsibility for the other, that motivates action. Elizabeth Segal (2007) defines this as going beyond the feeling-for or feeling-with an individual towards understanding the social and political structures of our society. This then is much more than "putting oneself in the other's shoes," but assumes responsibility for one's own role in somebody else's story. It creates urgency for practice, for action. This move from empathy to compassion is not an easy process and not always possible within contexts constrained by institutional requirements and limited resources.

We define compassion within the context of learning design along four dimensions: a desire to create more participative spaces, a recognition of power and positionality and how this affects our ability to participate, and finally a centreing of affect in the learning process. These three dimensions should then lead to a fourth dimension: the commitment to act towards more socially just learning design approaches, what we term *praxis* leaning on Paulo Freire's work, who argues that we need both "reflection and action upon the world in order to transform it" (1970, p. 51).

In this chapter, we will first introduce the four dimensions that constitute our compassionate learning design model. The model is based on a desire for more participation, an understanding of positionality and power dynamics in the classroom, the importance of affect, and how care and our mutual responsibility for each other lead to action. We then provide examples of how these principles could work together in practice. We use the terms "educator" and "learners" in a sector agnostic way, inclusive of school and post-school teaching and learning settings. Drawing on notions of teaching as a design science (Laurillard, 2012) we also use the term "designers" to refer to educators and Instructional or Learning Designers.

Towards Compassionate Learning Design: The Praxis of Participation, Justice, and Care

How does this work on empathy and compassion in the classroom relate to our work in learning design? In particular in traumatic times as we have experienced over the last two years? We are deeply concerned with designing for social justice, and we see compassion as coming from a place of care, a sense of justice, a desire to empower through increased participation levels of learners, and a compulsion to act collectively to recognise and act upon the emotional and mental health challenges and social need within our community in a reciprocal manner. However, we are also acutely aware that we live in highly unequal contexts and that power dynamics rule our classrooms, our relationships with other educators and with learners, and the relationships of learners among each other. As Sarah Sentiles (2017) states, quoting Judith Butler: "We are bound by what differentiates us as unique and irreplaceable and by our responsibility to others we don't understand. As Butler writes, 'Your story is never my story.' What's required is staying in relationship, even when we can find no common ground. Especially when we can find no common ground" (para. 7). This is why Zembylas' or Boler's writings on compassion when listening to stories of trauma, while they write from different contexts (teacher education in Cyprus and Canada), are so useful for us to think through what compassionate learning design would look like.

Recognising that we are different from our learners, and they may be different from each other is an important starting point, as Sarah Sentiles (2017) reminds us: "The challenge is to learn to live with, and protect, what we can't understand" (para. 6) It therefore seems essential not to assume that educators can fully know what learners want or need: "In the caring approach, we would prefer to advise: do unto others as they would have done unto them" (Noddings, 2012). Applied to education, we could say, "Do unto students as THEY would have done unto THEM" (Bali, 2021). But what does it mean to do to learners as they would want done unto them? How do we know what learners want done unto them, recognising our own positionality as educators in interpreting what learners need and how their needs might be addressed, and recognising the power dynamics among learners themselves?

We are influenced by several scholars who write on this. Firstly, Nancy Fraser (2005) developed the term "parity of participation", ensuring that everyone has equal power of decision-making when they come to the table. It means that "all the relevant social actors [...] participate as peers in social life" with an emphasis on the process involved "in fair and open processes of deliberation" (Fraser, 2005, p. 87). However, Fraser recognises that social injustice can occur across three dimensions: economic (access to resources), cultural (recognition), and political (representation). All three of these must be addressed in order for parity of participation to occur. Participation of "those who are intersectionally disadvantaged" is essential, along with recognition of their culture, "to ensure a more equitable distribution of design's benefits and burdens; fair and meaningful participation in design decisions; and recognition of community-based design traditions, knowledge, and practices" (Costanza-Chock, 2018). However:

> Our choices are deeply shaped by the structure of opportunities available to us so that a disadvantaged group comes to accept its status within the hierarchy as correct even when it involves a denial of opportunities… In turn, our agency and well-being are diminished rather than enhanced… Unequal social and political circumstances (both in matters of redistribution and recognition) lead to unequal chances and unequal capacities to choose. (Walker & Unterhalter, 2007, p. 6)

Therefore, simply creating a participatory space and inviting marginalised groups is unlikely to magically create equity, due to historical internalised and institutional oppressions. In order to achieve socially just participation, we must recognise that "justice needs care because justice requires the empathy of care in order to generate its principles" (White & Tronto, 2004, p. 427, citing Okin, 1990). We must also recognise variability in individuals' capacities: "the notion that one model of care will work for everyone is absurd…humans vary in their abilities to give and receive care" (White & Tronto, 2004, p. 450). If we strive towards democratic care, we need to place ourselves in the positions of both caregivers and receivers:

> [D]emocratic care requires switching perspectives and not just thinking about what we want. We need also to look at care from the standpoint of care-receivers, who will have different ideas about what kind of care they want or need to receive… In a "caring-with" democracy, we can set a goal of structuring institutions and practices so that each person's individual preferences can be honored. (Tronto, 2015, p. 34)

Therefore our understanding of compassionate learning design has four dimensions:

1. The desire to increase agency and participation of learners in their own learning process – PARTICIPATION
2. An understanding of power and history and how that affects our ability to participate: our positionality and intersectionality and how they influence our critical pedagogy – JUSTICE
3. A recognition of importance of affect and how that impacts learning: humanising, caring, and trauma-informed pedagogies – CARE
4. The aforementioned dimensions resulting in a commitment to act, to take responsibility and move towards more socially just learning design – PRAXIS

Figure 2 shows the relationship between care as overarching principle, participation (parity) as process, and justice as a desired (even if never reached) goal—with praxis as the result of these intersecting elements.

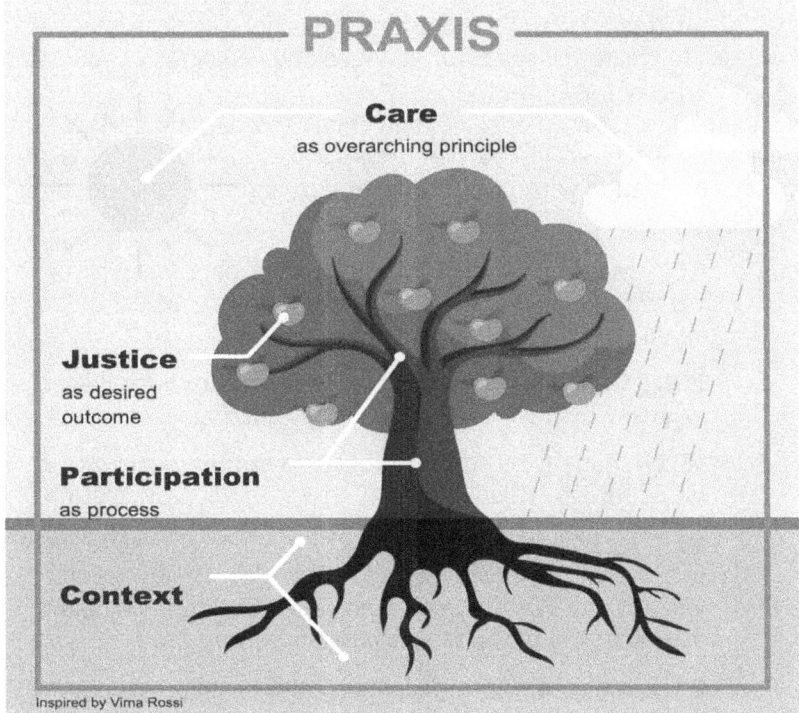

Figure 2: Towards compassionate learning design

Compassionate Learning Design as Participation, Justice, Care, and Praxis

Participation

> "[R]adical pedagogy must insist that everyone's presence is acknowledged...There must be an ongoing recognition that everyone influences the classroom dynamic, that everyone contributes" (bell hooks, 2014).

Humanising pedagogies draw from Paolo Freire (1970) and other critical pedagogues who describes humanising pedagogy as a revolutionary approach to teaching and learning that "ceases to be an instrument by which teachers can manipulate learner, but rather expresses the consciousness of the students themselves" (p. 51). Challenging the "banking system" of education, Freire advocates for learning that is based on problem-posing learning, dialogue, embodied learning, and democratic access to information to allow for transformative education. All of this needs the inclusion of learners in decision-making processes to create the space for dialogue and democrative education.

Wehipeihana's (2013) model of Indigenous evaluation can be helpful in showing different approaches to participation. Her model is about Western evaluation with Indigenous groups, and the levels of doing so involve:

- **TO**: evaluation done TO indigenous group, Western experts know best, and this is the most harmful form of evaluation
- **FOR**: evaluation done FOR Indigenous groups but by Westerners, which is benevolent but patronizing
- **WITH**: done together but probably with Western ways of doing things. This is the first step towards participation
- **BY**: done by and led by Indigenous groups (representation) but possibly still using world views of the West or need to explain ways of doing things—i.e. bringing participants to a table where the table is already set
- **AS**: led by Indigenous people and also complete autonomy to do so with their worldview and not having to justify—i.e. participants design their own table

Inspired by this model, we adapt it to compassionate learning design: we replace the Westerner/Indigenous groups with Educator/Learners and start from a place of benevolence where the educator wishes to act empathetically, and moves towards learner participation where compassion lives. These different levels refer to their process of deciding how to do so. We work with considering the needs and interests of actual learners rather than fictitious personas, since personas can become "objectified assumptions [which] then guide product development to fit stereotyped but unvalidated user needs" (Costanza-Chock, 2018). We ask how our design decisions impact on who/what is included and excluded? We move along two axes of increasing learner agency (political representation and parity of participation as in Fraser's work) in order to enhance learner empowerment and recognition (the cultural dimension of Fraser's work). Our adapted model would be represented by Figure 3 and described as follows:

- **TO:** educator knows and anticipates what learners need and offers one solution they believe will alleviate suffering or enable learners to learn best. This comes from a place of care but does not involve learner participation; it assumes educator's knowledge and expertise. More than anything, the educator prioritises the needs: they decide if the need for inclusion is paramount, or the need for economic security, or the need for something else, and because the educator has their own identity and positionality, they may not be fully aware of the spectrum of needs of learners, and which ones are a priority to tackle; they may make decisions about solutions that do not suit every learner.

- **FOR:** educator knows and anticipates what learners need, perhaps after asking them what they need via survey or asking in class, perhaps offering two solutions to choose from, thought of by the teacher independently without consulting with learners; it assumes educator has expertise. It is slightly better than the above because it involves some level of learner input, but again, the input from learners has to fit the boxes the educator has prioritised as important, and has decided can be open to some level of learner input, to the degree that the educator allows: learners make choices among options offered by the educator.

- **WITH:** educator and learners discuss needs and come up with a range of solutions together, but facilitated by the educator, probably within some institutionally-imposed restrictions. It involves nurturing learner agency, while maintaining overall power of setting parameters of negotiation in the hands of the educator. This can open the space for learners to bring in diverse needs and priorities not previously known to the educator, and to suggest strategies for addressing them not necessarily within the educator's existing arsenal of possible solutions.

- **BY:** learners organise themselves (with light facilitation perhaps) to discuss their needs, suggest a range of solutions, usually coming up with solutions similar to ones used by facilitators in the past, and need approval of the educator in the end in order for their solutions to be "accepted" within the institution for a degree or a grade or accreditation.

- **AS:** learners organise themselves without constraints of facilitator or institution, and do not need to justify their worldview, approach or choices in the end. This is learner self-determination. This could occur within an educational institution in some extra-curricular experiences or in community engagement projects, or outside of formal education altogether.

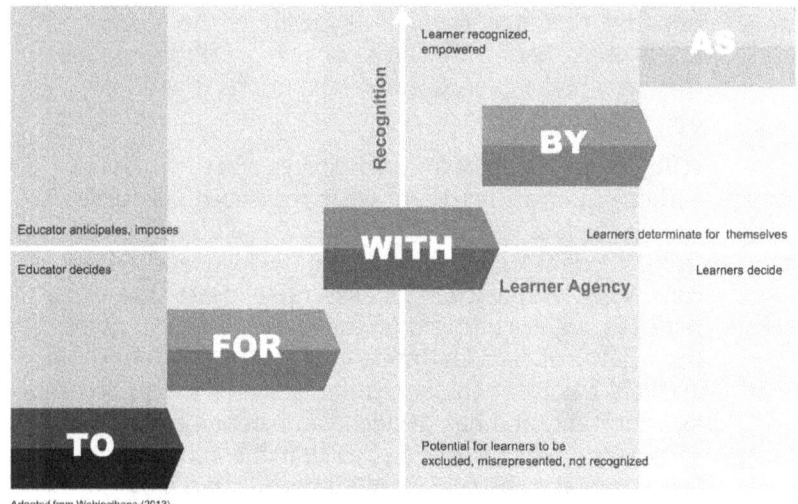

Figure 3: From designing with empathy to co-designing with compassion based on Wehipeihana (2013)

The following dimensions are present throughout the process, and reflective praxis around these considerations helps to enable moving from educator-led to learner-generated and from empathetic to more compassionate learning designs. There may be institutional constraints on an educator's ability to move towards more participatory levels in their practice, but it is important to enact as much empathy/compassion as possible within those constraints and to recognise the limitations within each level, while striving towards finding opportunities to enrich participation in areas where there is freedom to do so. Educators can also try to resist and advocate on an institutional level and form allyships for changing institutional policies and practices that go against caring, empathetic approaches, such as rigid grading policies.

Justice

Even as we move across these levels of self-determination of learners, we need to consider the unequal distribution of power between educators and learners and amongst learners themselves. We also need to question, to what extent, within formal educational contexts governed by institutional policies and external accreditation requirements, educators and learners can subvert and resist directives that reproduce injustice and reduce learner autonomy, as well as how historical constraints and oppressions have been internalised by some, and may influence their capacity to envision or express radical options. Moreover, as academic developers (or learning designers) supporting educators, we need to think of how encouraging and advocating for a compassionate design influences our process of working with educators who have different positionalities and teaching philosophies?

Is the role of the critical instructional/learning designer to advocate for particular pedagogical approaches based on their own values? Is their role to support educators implementing their own teaching philosophies? What if the educator's teaching philosophy and approach goes against the values of compassionate design, if care and social justice are not at the forefront of their minds? What if they have internalised beliefs that compassion goes against rigour, or what if institutions discourage compassionate approaches? As bell hooks reminds us, "Teachers who care, who serve their students, are usually at odds with the environments wherein we teach" (bell hooks, 2003, p. 91).

In practice, this may mean that, for example, we find the following scenarios:

An educator in a university may not have control over learning outcomes as they are predetermined by their department, but they can for example work "with" learners to decide on a range of assessments that can be used to meet those outcomes, and the rubrics may be created "by" learners working together. Underlying this praxis should be values of compassion and social justice, such that learners working together strive towards equitable processes and outcomes for each other and not just for their own parochial interests or ones they naturally empathise with based on their intersectional identities and past experiences.

In another situation, an educator teaching a graduate course may have freedom to co-create learning outcomes "with" learners, possibly having different groups of learners having different outcomes, and have learners work independently to develop their own individual assessments and assessment criteria ("as"), with feedback from the class community, but without requiring educator approval.

In both of these models, the educators' own positionality and intersectional identities and experiences of learners need to be considered, as any collaborative negotiation among learners is not inherently equitable, and learners unused to making choices and decisions are not necessarily well-prepared to participate in empowering ways (EquityXDesign, 2016). Sarah Sentiles writes: "The challenge is to learn to live with, and protect, what we can't understand… For the theorist Judith Butler, the ethical surfaces not when we think we know the most about each other, but when we have the courage to recognize the limits of what we know. Like Levinas, she proposes an ethical system based on difference, on relationships with others unlike you" (2017).

Activities that could be included in learning design processes that would allow educators and learners to reflect on their positionality are, for example, co-creating empathy maps and persona activities, not only for learners but also educators. If one maps the actual, not imagined, learner personas and educator personas, one might be able to see where overlaps are but also where gaps in understanding could occur. Other examples are Meta-Empathy Maps, which consider not only individual circumstances, but also institutional and structural conditions, that either include or exclude learners (EquityXDesign, 2016). Enriching these design activities with storytelling spaces allows for lived experience to surface and for a deepening of relationships between educators and learners and learners among themselves. Also raising awareness of the challenges of working across differences in co-design spaces is important and there are useful guidelines for working with and across hierarchies. In general, creating spaces for pause, for meta reflection, for preparing and supporting collaboration alongside the collaboration itself is recommended if one desires to create spaces that are truly participatory (see for example Ngoasheng et al, 2019 or Bradshaw, 2017).

Care

> "To teach in a manner that respects and cares for the souls of our students is essential if we are to provide the necessary conditions where learning can most deeply and intimately begin" (bell hooks, 2014).

Over the last year, approaches to teaching and learning that centre care and concerns for learners' well being have gained traction. Michelle Pacansky-Brock (2020) for example developed four principles for a humanising online pedagogy, that prioritises and serves connection and a feeling of belonging, "the connective tissue between students, engagement, and rigor" (ibid, p. 2):

- Trust: As educators, it is our responsibility to intentionally cultivate learner trust, and one way to do it is by practising "selective vulnerability" (Hammond, 2014 cited in Pacansky-Brock, 2020) in the online communities we build with our learners. As examples she mentions choosing to share aspects of our life that portray us as a real person—telling stories about a personal struggle we worked through or recording a video while cooking dinner or walking our dog.

- Presence involves intentional efforts to construct our authentic self through brief, imperfect videos to ensure our learners know we are in this journey with them (Costa, 2020). Verbal and nonverbal cues add context to our communications, which is particularly important to support culturally diverse learners.

- Awareness is achieved by learning about who our learners are and how we can support them.

- Empathy requires us to slow down, see things through our learners' eyes without judgement, be flexible, and support them towards their goals.

In similar fashion, Maha Bali (2020) advocates for a pedagogy of care, which allows us to get to know our learners and our learners to get to know us. This means making ourselves vulnerable, modelling sharing, so our learners become comfortable to share with us. As bell hooks suggests "empowerment cannot happen if we refuse to be vulnerable while encouraging learners to take risks." Bali calls this a hospitable environment, where everyone is given the space to choose whether and how to share. She also emphasises the importance of empathy with learners, while understanding and recognising that one can never fully understand what the other is going through. In Bali's model, educators enact relational care (Noddings, 2012) at various levels: At the course design/planning level, in habitual practices in the classroom, in the ways they respond to learners in the class environment, and finally, in their one-on-one interactions with learners. Crucial to this notion is combining both equity and care (Bali & Zamora, 2020; Bali

& Zamora, 2022a), manifesting in "democratic care" (Tronto, 2015), "parity of participation" (Fraser, 2005) and/or "Intentionally Equitable Hospitality (Bali et al, 2019; Bali & Zamora, 2022b). Some elements of this need to occur at an institutional level, or else the burden of affective care will fall on a few individuals, and the care will be selective and only ameliorative, rather than systemic or transformative. Short-term examples of institutional care can be seen in the establishment of "meeting free weeks" or institutional leave days for all employees (including academics and support staff). An example with more longer-term impacts of this in practice is how the University of Michigan Dearborn helped resist the use of remote proctoring technologies that reproduce inequalities and increase learner anxiety (see Silverman et al, 2021). They recognized that asking all educators to change their exams into alternative assessments would be a heavy burden of labor hours, and so they hired human graders to help grade these more complex assessments, instead of asking educators to accept the additional workload. Another example (see Bali & Zamora, 2022a) would be institutional support for learners with disabilities, where an allyship between a center for disability services, IT accessibility support, and teaching faculty exists so that individual educators did not have to scramble to find solutions for teaching learners with hearing or visual disabilities on a case by case basis.

Trauma-informed approaches to pedagogy have been used in higher education by educators with neuroscience and trauma and resilience training such as Mays Imad (2021, 2020) and Karen Costa (2020). Both draw on the Substance Abuse & Mental Health Services Administration (SAMHSA) definition of trauma and its six principles that guide a trauma-informed approach: safety, trustworthiness and transparency, peer support, collaboration and mutuality, empowerment and choice; and cultural, historical and gender issues. SAMHSA (2014) defines trauma as "an event, series of events, or set of circumstances that is experienced by an individual as physically or emotionally harmful or threatening and that has lasting adverse effects on the individual's functioning and physical, social, emotional, or spiritual well-being."

Mays Imad (2020) brings the notion of trauma into a Higher Education space and argues that trauma-informed pedagogy involves awareness of learners' past and present experiences, and how this impacts their well-being and ability to learn. Imad suggests that learning cannot happen when learners are dealing with trauma. She shows, from neuroscience research, how emotions influence learning. Therefore there needs to be space within the learning experience to engage and reflect on emotions. It is only "when our nervous system is calm, [that] we are able to engage socially, be productive, and process new information in order to continue to learn and grow—and to feel we are living meaningful and fulfilled lives" (p.2). She argues that in times of COVID-19, making space for engaging with trauma was particularly important, as both learners and staff experienced trauma, primary or secondary, which happens when learners and teachers witness each others' trauma and transfer trauma to others. Experiences of trauma include sickness and loss of family and friends, loss of employment, or "just" the experience of month-long isolation and loss of contact.

From our experience, before the pandemic, educators in general saw concerns around mental health and learner well-being not necessarily as integral to their own classroom practice, but rather as something to be "outsourced" to the university counselling centre and something they had little control over—they neither felt that their classrooms were spaces that could contribute to learners' anxiety or depression, nor as spaces that could help promote wellbeing of learners without necessarily becoming counselling or therapy sessions.

With the pandemic we started seeing more educators recognise trauma (our own and that of our learners) as part of their classroom practice and the recognition that how they design learning matters in terms of how to respond to this trauma, to ensure that learners feel safe, empowered and connected. We recognize that some disciplines lend themselves to more space for engaging emotions. However, checking in with learners about how they feel each day, being attuned to how we assign them work and how it makes them feel, and discussing how they feel when working on a difficult project are all practices that can happen in any course or any discipline. Educators can redesign their courses in ways that better enable learners to heal and succeed with the support of their community despite the trauma they experience.

Trauma-informed approaches aim at reducing uncertainty to foster a sense of safety, level communication to help forge trust, reaffirm or re-establish goals to create meaning, make intentional connections to cultivate community, and centre well-being and care. The word "trauma" has been used in a more collective sense as well in the pandemic context as a shared traumatic experience, given diverse positionalities we know that individuals have been affected differently.

Commitment to praxis (reflection and action)

Once these three dimensions (participation, justice, and care) are in place, we believe that the fourth dimension comes to play, i.e. where educators can commit to action, in our case to a more socially just learning design. Table 1 integrates Weihipehana's participation model with our elements of compassion learning design, to allow us to think through the various stages of participation, justice, and care. We are thinking through important questions or design considerations, such as who is the individual (or group) making key decisions, who has final approval power, what kind of conditions are conducive to this level, and examples for that level of participation and how justice and care would appear in each of them. This is thinking-in-process and we would appreciate feedback and further development by readers.

Conclusion

Compassionate learning design involves reflection of and attention to who makes choices in teaching and learning and the implications thereof. We argue that considering the intersecting dimensions of participation, justice, care, and praxis provides a helpful approach to learning design that is both compassionate and critical. In addition to contextual factors that constrain participation levels, designing for learner positionalities is a practice that educators, IDs/LDs, and co-designers of learning opportunities often do not fully understand. Also better understanding one's own positionality and how this drives particular designs over others is important. When designing in times of uncertainty and flexibility, we believe that educators and learners need to be conscious of intentionality and agility around cultivating compassion. We encourage readers to see this as a developmental and social process that is forever changing as you collaborate with fellow educators, IDs/LDs, and learners and grow your praxis over time.

	Participation				
	TO	FOR	WITH	BY	AS
Who takes decisions?	Educator takes design decisions based on anticipation / experiences from previous learners / courses (ie through a persona activity)	Educator takes decisions based on what is known from preliminary knowledge from actual learners (ie through surveys / focus groups))	Educator and learner take decisions together, facilitated by educator who sets constraints what possible (ie in a co-design process)	Learners take decisions with participation by lecturer, but are in charge, lecturer sets some constraints (student as partners)	Learners take decisions, process led by learners, educators not necessarily involved/do not approve (student-led designs)
Approval	Educator	Educator and Learners	Educator and Learners	Learners and Educators	Learners
Conditions	Educator constrained by institutional requirements, such as set curricula, accreditation requirements, unified assessments etc.	Educator constrained by institutional requirements, such as set curricula, accreditation requirements, unified assessments etc but educator has some agency in terms of setting teaching and learning activities, assessments etc.	Certain amount of flexibility and agency for educator to set parameters, such as learning objectives, curricula, T&L activities, assessments	High level of flexibility and agency for educator to set parameters, such as learning objectives, curricula, T&L activities, assessments	Complete freedom, learning happens outside institutional constraints

Table 1: Participation levels across Weihipehana's (2013) model adapted for compassionate learning design

Toward a Critical Instructional Design

	TO	FOR	WITH	BY	AS
Justice	Little reflection on power and hierarchies	Some reflection as educator responds to learners' feedback.	Reflection on power and hierarchies and how this affects collaboration between educator and learners is critical. Need for a parallel process to reflect on how power affects on participation.	Reflection on power and hierarchies and how this affects collaboration between educator and learners is critical. Need for a parallel process to reflect on how power affects participation	Focus on reflection on how power dynamics affect relationships between learners themselves. Educator's role is to lay the foundations for this but not be involved in the actual decisions
Care	Educator decides what learners may need and anticipates ways to care	Educator surveys learners, and this informs options the educator offers learners	Educator and learners openly discuss care needs and rights, and options	Learners drive the process to establish and implement strategies for care, educators approves, experiences of trauma integrated into curriculum	Learners establish strategies for caring for each other, experiences of trauma become the curriculum

Praxis (Examples)				
TO	FOR	WITH	BY	AS
Multisection mathematics course, unified assessments/curriculum, but educator takes account of of previous learner' struggles/needs and creates flexible deadlines and culturally relevant examples and authentic assessments (all educator pre-defined)	Multisection business course, but educator has some leeway over assessments but not learning outcomes. Surveys learners at the beginning of the semester and selects options of assessment based on learner interests and needs. Educator pre-defines different options and negotiates some deadlines and such with learners.	Elective course where educator has freedom over learning outcomes and assessments Educator always in negotiation with learners over learning outcomes and assessments BUT educator sets the "rules for engagement", facilitates the discussion, and makes final decisions. Educator can negotiate grading criteria and grades with learners, but ultimately holds power over final grades	Co-curricular community-based learning course with an educator as "advisor" but learners lead in terms of where to go, what to do, how to do it. Learners decide on grading criteria, but an educator still officially sets a grade	Extracurricular activity, completely learner-led Community engagement / Adult Education settings. Educators not involved in assigning a "grade" at all

References

Anvari, F., & Richards, D. (2018). Personas with knowledge and cognitive process : tools for teaching conceptual design. *Twenty-Second Pacific Asia Conference on Information Systems (PACIS), Proceedings*. 43. Japan: PACIS. https://aisel.aisnet.org/pacis2018/43

Bali, M. (2021). Do unto students as they would have done unto them. *Times Higher Education Campus*. Retrieved March 29, 2022 from https://www.timeshighereducation.com/campus/do-unto-students-they-would-have-done-them

Bali, M. (2020). Pedagogy of care: COVID-19 edition. *Reflecting Allowed*. Retrieved March 29, 2022 from https://blog.mahabali.me/educational-technology-2/pedagogy-of-care-covid-19-edition/

Bali, M., Caines, A., Hogues, R., Dewaard, H. & Christian, F. (2019). Intentionally equitable hospitality in hybrid video dialogue: The context of virtually connecting. *eLearn Magazine*. Retrieved March 29, 2022 from https://elearnmag.acm.org/archive.cfm?aid=3331173

Bali, M., & Zamora, M. (2020). Equitable emergence: Telling the story of #EquityUnbound in the open. *#OpenEd20 Plenary*. Retrieved March 29, 2022 from http://youtu.be/NEeZvM6_8UE

Bali, M. & Zamora, M. (2022) .The Equity-Care matrix: Theory and practice. *Italian Journal of Educational Technology*. https://ijet.itd.cnr.it/article/view/1241

Bali, M. & Zamora, M. (2022). Intentionally equitable hospitality as critical instructional design. In M. Burtis, S. Jhangiani, & J Quinn (Eds.), *Designing for care*. Hybrid Pedagogy. https://designingforcare.pressbooks.com/chapter/intentionally-equitable-hospitality-as-critical-instructional-design/

Boler, M. (1999). *Feeling power: Emotions and education*. New York: Routledge.

Borkoski, (2019). Cultivating belonging. *Equity & Access PreK-12*. Retrieved July 14, 2022 from https://www.ace-ed.org/cultivating-belonging/#:~:text=Who%20does%20it%20benefit%3F,Walton%20%26%20Carr%2C%202012.

Bradshaw, A. C. (2017). Critical pedagogy and educational technology. In A. D. Benson, R. Joseph & J.L Moore (Eds), *Culture, Learning and Technology* (pp. 8-27). New York: Routledge.

Costa, K. (2020). Trauma-Aware teaching checklist. *100 faculty*. Retrieved March 29, 2022 from https://bit.ly/traumachecklist

Costanza-Chock, S. (2018). Design justice: towards an intersectional feminist framework for design theory and practice. *Proceedings of the Design Research*

Society. Limerick: DRS. https://ssrn.com/abstract=3189696

Czerniewicz, L., Agherdien, N., Badenhorst, J., Belluigi, D., Chili, M., Villiers, M. de, Felix, A., Gachago, D., Ivala, E., Kramm, N., Madiba, M., Mistri, G., Mgqwashu, E., Pallitt, N., Prinsloo, P., Solomon, K., Strydom, S., Swanepoel, M., Waghid, F., & Wissing, G. (2020). A wake-up call : Equity , inequality and COVID-19 emergency remote teaching and learning. *Postdigital Science and Education*, 2(3), 946-967. https://link.springer.com/article/10.1007/s42438-020-00187-4?fbclid=IwAR3dWEIwpRz4r7ox3izGoj64g1Flee6eJToIDO-Qg42MuPkl1R7211mc7YoM

EquityXDesign (2016). Racism and inequity are products of design. They can be redesigned. *Medium*. Retrieved March 29, 2022 from https://medium.com/equity-design/racism-and-inequity-are-products-of-design-they-can-be-redesigned-12188363cc6a

Fraser, N. (2005). Reframing Justice in a globalized world. *New Left Review*, 36 (Nov/Dec). Retrieved March 29, 2022 from https://newleftreview.org/issues/ii36/articles/nancy-fraser-reframing-justice-in-a-globalizing-world

Freire, P. (1970/2005). *Pedagogy of the oppressed*. New York: The Continuum International Publishing Group.

hooks, b. (2003). *Teaching community: A pedagogy of hope*. New York: Routledge.

hooks, b. (2014). *Teaching to transgress*. New York: Routledge.

Imad, M. (2021). Transcending adversity: Trauma-Informed educational development. *To Improve the Academy*, 39(3). Retrieved March 29, 2022 from https://quod.lib.umich.edu/t/tia/17063888.0039.301?view=text;rgn=main

Imad, M (2020). Leveraging the neuroscience of now. *Inside Higher Ed*. Retrieved March 29, 2022 from https://www.insidehighered.com/advice/2020/06/03/seven-recommendations-helping-students-thrive-times-trauma

Gachago, D., Van Zyl, I. & Waghid, F. (2021). More than delivery: Designing blended learning spaces with and for academic staff. In Sosibo, L. & Ivala, E. (Eds.), *Transforming learning spaces* (pp. 132-146). Vernon Press.

Laurillard, D. (2012). *Teaching as a design science: Building pedagogical patterns for learning and technology*. Routledge.

Ngoasheng, A., Cupido, X., Oyekola, S., Gachago, D., Mpofu, A., & Mbekela, Y. (2019). Advancing democratic values in higher education through open curriculum co-creation: Towards an epistemology of uncertainty. In L. Quinn (Ed), *Reimaging curricula: spaces for disruption*, p.324-44. African Sun Media. https://doi.org/10.1093/0198294719.001.0001

Noddings, N. (2012). The language of care ethics. *Knowledge Quest*, 40(5), 52-56.

Pacansky-Brock, M. (2020). How and why to humanize your online course. Retrieved March 29, 2022 from https://brocansky.com/humanizing/infographic2

Segal, E. (2007). Social empathy: A tool to address the contradiction of working but still poor. *Families in Society*, 88 (3), 333-37.

Sentiles, S. (2017). We're going to need more than empathy: We have to get radical with the idea of the other. *LitHub*. Retrieved March 29, 2022 from https://lithub.com/were-going-to-need-more-than-empathy/

Silverman, S., Caines, A, Casey, C., Garcia de Hurtado, B., Riviere, J., Sintjago, A., & Vecchiola, C. (2021). What happens when you close the door on remote proctoring? Moving toward authentic assessments with a people-centered approach. *To Improve the Academy*, 39(3). https://doi.org/10.3998/tia.17063888.0039.308

Substance Abuse and Mental Health Services Administration (SAMHSA) (2014). *SAMHSA's concept of trauma and guidance for a trauma-informed approach.* (HHS Publication No. (SMA) 14-4884). US Department of Health and Human Services. Retrieved March 29, 2022 from https://ncsacw.acf.hhs.gov/userfiles/files/SAMHSA_Trauma.pdf

Tronto, J. C. (2015). *Who cares?: How to reshape a democratic politics.* Cornell University Press.

van Zyl, I. & de la Harpe, R. (2014) . Mobile application design for health intermediaries. Considerations for information access and use. *BIOSTEC 2014: Proceedings of the International Joint Conference on Biomedical Engineering Systems and Technologies*, 5, 323-328. https://doi.org/10.5220/0004800803230328

Walker, M., & Unterhalter, E. (2007). *Amartya Sen's capability approach and social justice in education.* Palgrave Macmillan.

Wehipeihana, N. (2013). *A vision for indigenous evaluation* [Keynote conference presentation]. Australasian Evaluation Society Conference, Brisbane, Australia. https://youtu.be/H6LXD3RjqLU

White, J. A., & Tronto, J.C. (2004). Political practices of care: Needs and rights. *Ratio Juris*, 17(4), 425-453.

Zembylas, M. (2008). Engaging with issues of cultural diversity and discrimination through critical emotional reflexivity in online learning. *Adult Educa* tion, 59 (1), 61–82. http://adlawrence.blogs.wm.edu/files/2011/03/cultural_diversity_emotional_reflex.pdf

—. (2011). *The politics of trauma in education.* Palgrave Macmillan.

Indigenizing Design for Online Learning in Indigenous Teacher Education

Johanna Sam, Jan Hare, Cynthia Nicol, and LeAnne Petherick

How do you bring Indigenous knowledges into learning management systems (LMS)? How do you weave Indigenous perspectives in the course design while using a LMS that can be seen as dominant/Eurocentric? Indigenous Teacher Education Programs (ITEPs) play a critical role in preparing Indigenous teacher candidates (ITCs) to serve Indigenous learners, schools, and communities. While ITEPs may allow Indigenous students to remain in their communities for their teacher education programming, ITEPs are generally part of mainstream teacher education. Critiques of teacher education have established how classrooms operate as colonized spaces (Cote-Meek, 2014). As a result, dominant/Eurocentric theories and practices of teacher education curriculum tend to marginalize ITCs' knowledge and experience in coursework (Brayboy & Maughn, 2009). ITEPs need to innovate to create distinctive curriculum and pedagogies that prepare ITCs for blended classrooms (e.g., online and in person teaching) and Indigenous communities (Hare, 2021).

As Indigenous and ally teacher educators and scholars, we come together to promote an Indigenizing design approach for teacher education curriculum that centers Indigenous perspectives, histories, worldviews, and pedagogies in online learning environments in respectful and productive ways. Central to our work is validating ITCs' experiences as legitimate knowledge sources in coursework.

While there is growing scholarship that describes ITCs learning from Indigenous knowledges (Brayboy & Maughan, 2009; Garcia & Shirley, 2013; Whitinui, Rodriguez de France, & McIvor, 2018), we situate our collaborative curriculum design process within new multi-modalities that bring Indigenous knowledges and pedagogies into online learning spaces. We share pedagogical principles guiding Indigenization of a set of courses that are part of the curriculum in the professional certification year of the Bachelor of Education program for ITCs in NITEP - the Faculty of Education's Indigenous Teacher Education Program at the University of British Columbia (UBC) in western Canada. We offer personal and professional insights on the application of these principles, highlighting the Indigenizing design process, providing examples of how Indigenous knowledges and pedagogies operate within and beyond the digital boundaries of synchronous and asynchronous coursework, and reflections on student engagement and learning.

In line with Indigenous protocols to knowledge, we situate ourselves in relationship to the matters on which we write. We are a collective of scholar-educators who work in a large teacher education program in western Canada. Two of us are Indigenous. Johanna is a proud citizen of Tŝilhqot'in Nation in north-central British Columbia. Johanna's research and teaching takes a strength-based approach for exploring digital spaces and wellbeing among youth and Indigenous communities. Jan is an Anishinaabekwe from the M'Chigeeng First Nation in northern Ontario. Her teaching and research are concerned with how Indigenous and ally teachers respond to Indigenous knowledges and pedagogies in their classroom practices. In her role as Director of NITEP, she has sought to transform mainstream teacher education programming to be more responsive to ITCs. LeAnne is a fourth-generation settler scholar with Scottish and English heritage, who grew up on the traditional Lands of the Anishinabewki, Mississauga, and Wendake-Niowentsio. Her teaching and research are grounded in social justice issues in Physical Education, Health Education, Sport and Teacher Education. She is committed to social justice issues as a strategy for shifting the dialogue, practice, and experience of people through human movement practice. Cynthia is a seventh-generation settler Canadian of German and English ancestry raised on the Ktunaxa (Kootenay) territory in southern British Columbia and learned to teach mathematics on Haida Gwaii in BC's Pacific northwest coast. Cynthia's research and teaching focuses on working with communities and teachers bringing together mathematics, community, culture, and place.

Indigenizing Learning Design

For us, Indigenizing design is not about replacing the Eurocentric curriculum of teacher education with Indigenous content. Rather, it is about embedding Indigenous perspectives and histories across critical dimensions of course design that includes objectives, learning activities, assessments, and pedagogies to form a "plan for learning" (Thijs & van den Akker, 2009). It is a process that is responsive to local lands, languages, traditions, and knowledges where ITCs live and learn. It reflects a commitment to naturalizing Indigenous ways of knowing within learning design. Pete, Schneider, and O'Reilly (2013) suggest

Indigenizing is about resistance to the colonizing tendency to erase Indigenous people and the persistence of Indigenous ways of knowing. Grafton and Melancon (2020) explain Indigenization as "a process of resurgence, a recentring of precolonial and colonial ways of knowing and being that never ceased to exist despite colonial structures and process and their attempts to assimilation and erasure" (p. 142). Our Indigenizing frame of reference focuses on online teaching and learning that is inclusive of Indigenous knowledges and pedagogies in ways that empower ITCs and the school and communities they serve.

Digital environments are not neutral spaces, nor should they be considered landless (Geartner, 2016). Morford and Ansloos (2021) tell us that "with the rise of computer-based technology, settler colonialism has seeped into the cyber-realm. The Internet has become another space and place where the violence and normalization of colonization are perpetuated" (p. 295). While efforts to represent Indigenous knowledges in online spaces risks appropriation, misrepresentation, or commodification, Wemigwans (2018) believes in the transformative potential of Indigenous knowledges in digital spaces to contribute to Indigenous healing and resurgence when Indigenous communities control information that is generated and shared online. Similarly, Article 14 of the United Nations of Declaration of Rights for Indigenous Peoples, indicates that "Indigenous peoples have the right to establish and control their educational systems." With respect to providing education in a manner appropriate to Indigenous teaching and learning, we focus on the ways Indigenous knowledges and pedagogies can be organized, mediated, and practiced within and beyond the online learning space to promote Indigenous values in virtual environments (de Haan, 2018).

Indigenous Teacher Education: A Context for Digital Design

NITEP is a well-established Indigenous teacher education program that prepares ITCs for education roles in classrooms, schools, and communities. It offers community-based programming in remote, First Nations, and rural areas of the province of British Columbia, Canada, allowing ITCs to remain at a local field center for the first 3 to 4 years of the program before transitioning to UBC's Vancouver urban campus to complete their professional certification year of their 5-year concurrent Bachelor of Education degree. This move to campus results in student attrition and impacts on program completion due to family and community commitments, financial barriers, and cultural priorities experienced by ITCs. Review of research suggests that students prefer community-based approaches to teacher education, whereby they can become certified teachers without having to leave their home communities and territories (Whitinui, Rodriguez de France, & McIvor, 2018). In addition, flexible delivery modes, that include online learning, can increase access, participation, and successful completion of coursework for ITCs.

With support from the federal government ministry, Indigenous Services Canada, NITEP engaged in an Indigenizing approach to redesign a set of required courses in the teacher education program to an online delivery mode. These courses were designed and taught by Indigenous and non-Indigenous teacher educators. While these courses were redesigned for a group of 15 ITCs at a NITEP regional field center in northern British Columbia, they are intended for on-going delivery with other NITEP field centers and within the on-campus teacher education program. Those involved in the revisioning of courses were committed to Indigenous education and met regularly over two years to discuss design strategies and principles across varied content areas. Instructors of the courses also met regularly as a collective to support one another in teaching their courses, sharing and appraising resources, activities, and content and generating questions concerning cultural knowledge and local context of ITCs. Though learning through online modes may not have been the primary choice for ITCs, and neither were the courses intended to be fully delivered in digital spaces, the global pandemic necessitated conditions for synchronous and asynchronous instruction.

Pedagogical Principles for Indigenizing Design

To guide the development of online instruction and learning environments for ITCs, we formulated a set of pedagogical principles for Indigenizing design. We draw on both scholarship and experience to define the following four pedagogical principles:

Indigenous knowledge frameworks

Given the rich and diverse ways Indigenous knowledges are understood, it was helpful for instructors to utilize Indigenous knowledge frameworks to assist ITCs in applying Indigenous theory or concepts to their teaching practices. Within an Indigenous knowledge framework, "there are many ways by which knowledge can be organized...including taxonomy, ceremony, art, and ritual" (Varghese & Crawford, 2021, p. 10). Indigenous worldviews are expressed within these frameworks and knowledge, concepts, or values are then classified, integrated, or structured to provide an understanding of the world around us. Indigenous knowledge frameworks in their conceptual or visual representations demonstrate the sophistication of Indigenous knowledge systems, which often face challenges of validity in their application to academic disciplines. There are a growing number of Indigenous knowledge frameworks that assist with meaningful engagement of Indigenous theories and research (e.g., see Archibald, 2008; Kirkness & Barnhard, 1991; Styres, 2017; Wemigwans, 2018) and can be applied to instructional design.

Localization

Though Indigenizing design is attentive to the diversity in languages, cultures, and practices among Indigenous groups, there are common elements that were considered in adapting the mainstream teacher education curriculum to local contexts. For example, there are common values and pedagogies among Indigenous people that include oral tradition, intergenerational approaches, land and experiential learning, and relationality and interconnectedness. Indigenous scholarship assisted instructors to be inclusive of the broader values and pedagogies for Indigenizing, drawing on Indigenous scholars, educators, authors, or artists to contribute a broad range of Indigenous perspectives in the course design. However, responding to local contexts required instructors to consult and collaborate with local community members, knowledge keepers, and students. Co-author Johanna was from the local territory and provided guidance to the instructors. In some instances, ITCs were hired to support some aspects of local course development, given their prior experience with curriculum development.

Engaging Indigenous Elders, knowledge keepers, and community members in Indigenizing course design resulted in richer and fulsome experiences for both students and instructors. Local stories, knowledge of land markers, histories of place, languages, and interactions of students with community, family, and practicing educators amplified the local aspects in the instructional design. This required instructors to consider local protocols, which are systematic rules of acquiring and utilizing Indigenous knowledges in the course. For instance, cultural protocols may include how Indigenous knowledges and oral traditions are shared in social, political, educational, and cultural ways. While many of the instructors were sensitive to Indigenous protocols, community and students assisted instructors with this element of design.

Multimodalities

In the online space, instructors were able to apply a range of multimodal forms including audio, images, texts, or videos. For instance, digital storytelling is an important mode of expression and pedagogy for Indigenous people and communities to restore, generate, document, and archive their own histories, truths, and contemporary realities (Sam, Schmeisser, & Hare, 2021). Stories in digital forms have been described as "living breath" connecting learners to their ancestors, homelands, languages, teachings, and future generations (Manuelito, 2015). Multimodality in course design enabled ITCs to share historical and contemporary cultural visual expressions, listen and create podcasts, take part in traditional knowledge activities, such as drumming, songs, rattle making, preparing traditional foods, learning about traditional plants and medicines, and listening to and observing Elders and community members who took part as online guests. Instructors drew on digital tools, learning platforms, and social media to create opportunities for ITCs to utilize the growing number of cultural and language apps and LMS platforms as modes of digital sovereignty.

Design for relationship

Relationality configures strongly within Indigenous worldviews. This includes relationships to one another, to family, community, ancestors, and to land and place. Instructors placed significance on fostering and sustaining these relationships as a pedagogical principle. Despite being geographically dispersed, online learning can still create opportunities for holistic connections among learners. Reedy (2019) found a focus on design, creation, and facilitation of online spaces that prioritizes the relational nature of learning appealed to Indigenous students. This also involved relationships with instructors, where a strong instructor presence enhanced Indigenous students' feelings of connection, especially when the, "teacher exhibited Indigenous cultural awareness, such as taking flexible approaches to assessment time frames to enable students to balance their studies with their family and cultural obligations" (p. 141). Reedy suggests that when designing for Indigenous students, educators ensure ample opportunities for all students to make interpersonal connections and provide purposeful tools, spaces, and pedagogies to facilitate these interactions.

For instructors, an Indigenizing approach to online learning also emphasized relationships to the land. Place-based conceptualizations within digital spaces can occur through different mediums, such as messaging apps, Facebook, Twitter, Instagram, Tik Tok, or YouTube. Morford and Ansloos (2021) analyze social media environments to examine Indigenous conceptions of land in these spaces. With a focus on Twitter, these authors surmise that "tweets become a way of digitally rematriating settler occupied lands, asserting and reclaiming them as Indigenous land imbued with Indigenous spirituality, knowledges, and relationality" (p. 297). Online learning offers possibilities for ITCs to connect with lands and territories for instructional design. These spaces are living spaces where land- and place-based relationships can be developed for students.

Indigenizing Design Exemplars

Indigenous voices in educational psychology and special education courses

The developmental literature in educational psychology often focuses on individual differences (Adams et al. 2015). Further, the field of educational psychology continues to use standardized tests and diagnostic assessment, which can be entrenched in cultural bias (Skiba, Knesting, & Bush, 2002). As such, four Educational Psychology and Special Education (EPSE) courses were re-designed with emphasis on the pedagogical principles of localizing Indigenous perspectives and digital multimodalities. The curriculum design aimed to shift the focus towards a strength-based perspective to emphasize Indigenous voices, knowledges, and traditions in regards to human development, learning assessment, and cultivating supportive school environments. The exemplar presents Indigenous voices in multimodal resources within LMS to shift from the dominant/Eurocentric individual perspective to the inclusion of Elders and knowledge keepers' voices that express a "cultivation of collective well-being" (Adams et al. 2015, p. 221). The curriculum development gathered Indigenous voices in EPSE courses, which are shared in this exemplar.

To localize the curriculum, I (Johanna) started by consulting with faculty course coordinators, Indigenous community partners, and educational technology instructional designers. From these consultations, I established themed modules for the course. Multimodal activities within a LMS modules were asynchronous, engaging ITCs in critical reflections, peer discussions, and digital storytelling projects. ITCs were critical of the LMS in their consideration of content development when interviewing Knowledge Keepers and functions of LMS to deliver Indigenous digital resource repositories. LMS played the role of content delivery (e.g., course administration, learning resources, peer assessment) and a tool used to communicate in an online environment (e.g., announcements, messages, discussion forms). Two ITCs were hired to support local engagement with community members in the creation of digital multimodal resources. We collaborated together to interview Elders, Knowledge Keepers, and educators from the Tŝilhqot'in Nation and Nuxalk territories to create multimodal resources for the courses. The result was five short length videos that enhanced Indigenous voices in the learning design.

The ITCs and I co-developed an interview guide for the filming projects. ITCs identified individuals to interview in their respective communities. Elders and Knowledge Keepers provided their consent and permission to be interviewed. Questions posed to Elders focused on, *what does it mean to live a healthy (good) life?* Questions for teachers and school principals focused on, *how do you create a strong community?* Knowledge Keepers were asked, *how does storytelling or songs shape teaching and learning?*

Selected quotes from each interview are included to convey localization of Indigenous voices in the multimodal resources for the EPSE curriculum. Topics discussed in the interviews related to social and emotional development as well as cultivating supportive school and classroom environments. A Nuxalk Elder spoke about the importance of sacred knowledges.

> There are medicines that have to be developed in different ways that I was taught. And also, what I was taught is that some of the medicines we only share with very, very chosen people that will carry on the ways of the medicines with strict confidence.
>
> – E.V.Sun-Hwrna Schooner, Nuxalk Elder

In the next quote, an Indigenous school administrator who lives in the Nuxalk territory provided guidance to ITCs on how to cultivate positive relationships with students. She was asked, how do you build positive relationships with students?

> Building relationships, your consistency. Your consistency of your attendance. Of them knowing you're here for the long-

> haul. I would say recognizing emotions. Really, really talking through and validating to make sure kids understand that you know what they are going through and it's okay to have these big emotions. It's definitely a long-term thing. I would say like some of the relationships I have built with kids it took years to get that close. Now, they're comfortable walking in my door and that's good!
>
> – Desireé Danielson, Vice Principal Acwsalcta School

Then, a Tŝilhqot'in educator shared about the importance of building a sense of family among students, teachers, school administration, and staff for a supportive school environment. He was asked, what makes a strong learning community?

> A strong community at school, first of all, you have to pretty much develop a family setting where the students come to school and feel comfortable and get along with everyone. Not to have a stress-related environment. You pretty much have to get along with everyone, staff, parents, and children. The idea is to bring back the family sense, so that when they go to school, they are comfortable in that setting. And also, you want to achieve something that they look forward to on a daily basis, so that's what you want to develop over time. By doing this, you have to communicate with them.
>
> – Grant Alphonse, Tŝilhqot'in Nation

Then, a Tŝilhqot'in Elder spoke about the role of grief and loss among young people. She states the importance of cultural protocols in social and emotional wellness when experiencing the stages of grief. She was asked, what stories do you know about living a good life?

> I realized a lot of the young people don't know the protocol at all. That sorta really sticks out. And being a widow, hardly nobody knows the different stages they go through. Like with some of the Elders or with somebody that's been sick a long time and you know it's okay, but if they die suddenly that's when it's more devastating. That's what I notice, you go through a lot of stages. On top of that, when you're a widow, you got to follow your protocol. And with grief and dealing with everything, it's a really tough year. But a lot of what I notice is that a lot of young people come to me and ask, "how are you dealing with it?" And I tell them the different stage I am at.

– Agnes Alphonse, Tŝilhqot'in Elder

Interviews with Elders and educators conducted by ITCs assisted with the Indigenizing EPSE curriculum by engaging in local Indigenous stories, cultures, and practices. ITCs adhered to cultural protocols during the interview process. The curriculum development for the four EPSE courses enhanced Indigenous voices in a series of local community interviews to form multimodal resources that included video, audio, and text. These multimodal resources promoted relationship building among ITCs during their EPSE coursework via ongoing reflections and dialogue with peers based on the teachings presented in the interviews.

Mathematics on the land: Listening to the land and living rhythms

"Run Bella Run! Today is Math!" Our mathematics curriculum and pedagogy course began with one ITC voicing the sentiments of others when she posted her experience and relationship with math in this 6-word poem. Bella's experiences, like many, describe mathematics as a subject to be feared, only done in math class, and disconnected from personal, cultural, and everyday lives. This isn't surprising because, as Bishop (1990) argues, mathematics is a "secret weapon of cultural imperialism" considered as "one of the most powerful weapons in the imposition of Western culture" (p. 51). Indigenizing mathematics education and designing curriculum for ITCs required engaging with Indigenous design principles of relationality, Indigenous frameworks, localization, and multimodalities. Drawing upon Indigenous scholars and writers the course was rooted in five themes that focused on the importance of relationship, land, community, culture, and story (Archibald, 2008; Armstrong, 1998; Cajete, 2004; Coulthard, 2010; Ghostkeeper in Surkan, 2018; Kimmerer, 2013; Salmon, 2017; Simpson, 2017; Styres, 2017; Wagamese, 2019) with mathematics threaded across the themes (see Fig. 1).

As mathematics is often an emotional and sometimes traumatic experience for many, we began the course (re)building and (re)storying relations with mathematics through shared stories of mathematical experiences and math-walks on the land. ITCs collected and posted images and sounds of mathematical patterns and rhythms they noticed around them outside, on the land, and in their homes. We pushed the boundaries of what counted as mathematical by collecting images of symmetries, fractals, spirals, and sounds in both the human-built

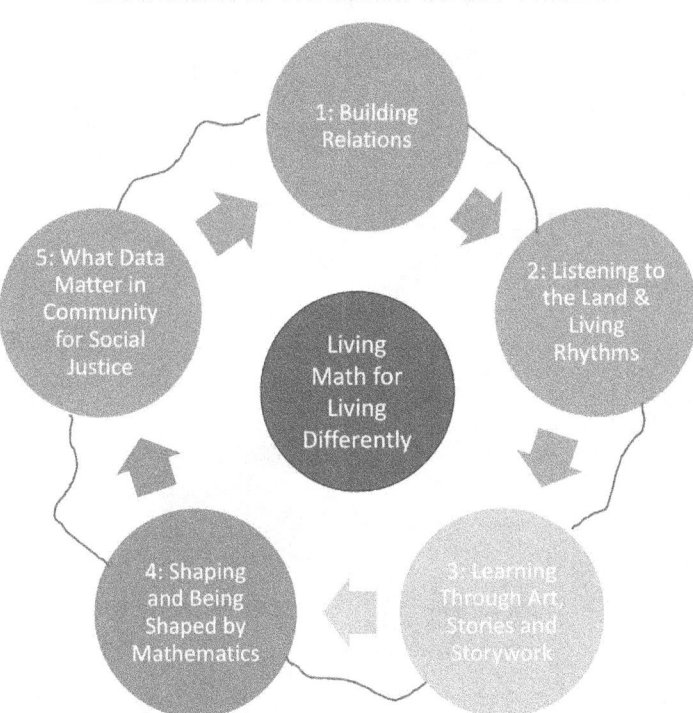

Figure 1: Elementary math education course themes

and more-than-human worlds. We posted our findings to the course Padlet (Fig. 2) to (re)story mathematical relations and to encourage ITCs in (re)membering (Styres, 2017) that listening to the land and its patterns can bring us together even across geographic locations. Focusing on local relationships but shared and distributed digitally allowed us to explore mathematics within our own contexts and notice pattern connections: the spiral of a spider's web in one ITC's image was connected mathematically to the spiral in a mountain goat's horns, the spirals in cauliflower, and to pine needle weaving of others' images. We began building relationships with each other, our communities, and mathematics through ITCs' shared knowledges, experiences, noticings, and stories.

For ITCs the task opened-up opportunities to acknowledge how we live mathematical actions such as noticing and studying patterns visually, orally, and with our bodies. We became each other's teachers through sharing stories that centered Indigenous pedagogies and mathematical conversations/inquiry while at the same time decentering mathemat-

Figure 2: ITCs posted images of mathematical patterns on the land

ics' power to define who can do mathematics and who is seen successful. A series of weekly math-walks kept this relational thread alive. A math-walk noticing what comes in 2s, 3s, 4s, or 5s had one student offer how blackberry bush leaves grow in clusters of 5 sparking curiosity of others to examine berry bushes in their areas. Does this occur for all berry bushes, or just black berries, or just black berries in your area? What else grows in 5s? Another math-walk connected to ITCs' previous contributions to developing educational resources for a digital storytelling project (Sam, Schmeisser, & Hare, 2021). This math-walk invited ITCs to explore the possible trading and economic practices along a 450 km exchange corridor known as the Nuxalk-Carrier Grease Trail, through mountain passes, forests and open meadows where westcoast ooligan fish grease was traded for blankets and tools. This math-walk offered possibilities to explore Indigenous wealth distribution, circular economies, and trading practices. A further math-walk on noticing pathways, both human and non-human, resulted in students sharing mathematical patterns of animal trails found on their mountain hikes and in gardens. Okanagan poet and scholar Jeannette Armstrong (1998) writes of the lifeforce of land that "holds all knowledge of life and death and is a constant teacher...the land constantly speaks. It is constantly communicating" (p. 176). For Armstrong land is a source of language and story. A goal of this course was to support ITCs in re-imagining mathematics as part of that story.

Stories formed a strong relational thread through the course with Ojibwe author Richard Wagamese (2019) writing that "it all begins with story" (p. 32). The course involved ITCs in exploring mathematics through story and building on the potential of story to connect mathematics and community. Stó:lō scholar Jo-ann Archibald Q'um Q'um Xiiem writes about the power of story for educating not only the mind, but also the spirit, body, and heart in her book Indigenous Storywork. ITCs were invited to inspire mathematical curiosity through experienced and embodied connections to community, specifically through story in digital, oral, and written modes. We drew upon Archibald's (2008) Indigenous storywork principles of respect, reverence, reciprocity, and responsibility in preparing to select and design a mathematics lesson with story, to become, using Archibald's words, "storywork-ready" (Indigenousstorywork.com). The principles of responsibility and reciprocity were particularly useful in considering stories with mathematics that could be specifically connected to, and give back to, community. Students worked in teams forming digital relationships facilitating collaborative lesson design across geographic areas.

Not all groups chose stories connected to land or culture, and no group chose oral stories. Some for example chose popular North American children's stories such as American writer Eric Carle's *The Very Hungry Caterpillar* where the mathematical connections were more explicit and the story perhaps more familiar and readily available during pandemic restrictions. Others chose stories by Indigenous authors including Dene writer Richard van Camp's book *What's the Most Beautiful Thing You Know About Horses?* Or Sakaw Cree writer Dale Auger's *Mwâkwa Listens to the Loon* in which one ITC engaged her students in discussions of Water Beings fishing to provide food for Elders and families. One group drew upon a local cultural resource, *Coyote's Food Medicine* written by Elders of Northern Secwepemc territory to design and try out lessons around mathematical patterns of seasons, gathering food, and cooking with family members within their COVID bubbles. This land-based mathematics education course brought us together in digital spaces to explore and be inspired by the study of patterns in our own land-based contexts. This provided opportunities for us to (re)story relationships with the land and mathematics, to experience the land mathematically, and to learn to live mathematics differently.

Centering land in physical and health education

The United Nations Declaration on the Rights of Indigenous Peoples states that "all peoples contribute to the diversity and richness of civilizations and cultures" (p. 2) and calls for an urgent need to reaffirm the fundamental importance of Indigenous self-determination. The call for Indigenous self-determination can be applied to Physical and Health Education (PHE), which emphasizes a lifelong approach to wellness. However, fostering Indigenous self-determination in PHE is not straightforward as there are both convergences and tensions between Indigenous and Western paradigms. For example, with its emphasis on embodied and experiential education, PHE aligns well with some of the core principles of an Indigenous paradigm (Styres, 2017; Simpson, 2014). At the same time, however, PHE has its roots in Euro-Western assumptions about movement as a decidedly individual and human-centered engagement. This individualized and anthropocentric framing is inconsistent with an Indigenous paradigm, which situates self-in-relation (Graveline, 1998) to human, more-than-human, Land, and spiritual relationships. Thus, when envisioning a course in PHE as a possible site for affirming Indigenous physical activity and health practices, the task was to attend to both the convergences and tensions at play. In doing so, Indigenous knowledges, ways of being and ITC's experiences in PHE were centered as a pathway for cultivating different experiences for children in elementary physical and health education, which became the overarching framework for the pedagogical principles, course development, and course administration.

The learning management system enabled ITCs to gather and share their stories in (re)visioning a PHE curriculum that centered Indigenous knowledges and pedagogical principles. This was facilitated through drawing upon Mohawk scholar, Sandra Styres' (2017) circular, four directional approach to Land-based education, the course moved through the following phases:

1. (Re) Centring Indigenous Ways of Knowing and Being in Physical and Health Education
2. (Re) Membering Indigenous Values of Relationships: Land, People, and Place
3. (Re) Generating Knowledge Systems through Activities, Sport, and Everyday Life

4. (Re) Actualizing Movement and Place-Based Knowledge Systems for Health

Both the four phases and the circular approach provided a framework that illuminated and centered the notion of self-in-relation to the knowledges, experiences, and communities, thus weaving together aspects of locality, which became the foundation of wellbeing for the course. Such an approach means that everyone in the course shared in a pedagogy that drew upon the momentum of the circle in a pendular fashion, constantly moving between phases as necessary, as opposed to a linear progression that centers pre-determined learning outcomes.

By foregrounding the experiential, kinesthetic, and affective relationships people have with local places, human movement and wellbeing, the cultural- and place-based specificity of Styres' approach created a space online for ITCs to build a connection between traditional and contemporary practices of Indigenous wellbeing centered around their own experiences, histories, and local knowledges. In so doing, the modalities used in the digital space created a pathway to advance a sense of reciprocal relationality (Wilson, 2012) whereby the ITCs knowledge and experience shaped the course. To this end, the course specifically allowed for emergent opportunities among ITCs to identify and share the interwoven connections with Land, culture, tradition, community, physical activity, health and wellbeing.

To give a few examples, in phase one, *(Re) Centring Indigenous Ways of Knowing and Being in Physical and Health Education*, ITCs were invited to share an artefact that was meaningful to them that reflected their connections with physical activity, health, wellbeing, culture, tradition, or community. During the first synchronous class, ITCs shared their artefact with the larger group, and this spontaneously generated additional sharing, where participants used the chat space to offer up a number of other related resources, such as photos, videos, and maps. In so doing, the online platform became a generative and lively space where story, place, values, and experiences proliferated in unanticipated, student-centered and holistic directions. In the second phase, *(Re) Membering Indigenous Values and Relationships*, a pivotal moment for relational understanding emerged. ITCs were asked to share their favorite flavor of ice cream. Almost instantly gooseberry ice cream, which was referred to as "Indian ice cream," was mentioned. The mention of Indian ice cream generated considerable excitement as ITCs voluntarily and enthusiastically shared their experiences with gooseberry ice cream. As an outsider unfamiliar with Indian ice cream, the ITCs exercised their insider knowledge, explaining to me the social, cultural, and place-based significance this particular food had for them. The class dialogue that ensued involved people sharing stories ranging from the physical activity necessary to locate, pick and preserve the berries, prepare the ice cream, all the way to sharing the desert with family and friends. It was clear that the experiential aspects of gooseberry picking were then layered with meanings about the ice cream's texture, taste, and use. As stories were verbally shared images popped up in the chat room, photographs of berries and family were offered, links were made to YouTube, and maps of where to locate bushes were identified. Using stories and technology it quickly became apparent that every ITC in the course had a relationship with gooseberry ice cream. As a settler scholar I was both excited and fascinated by the details relayed and I remarked about how each person in the course had a relationship with this food. It was at this time, that one of the ITCs said, "of course LeAnne, it's Indian ice cream!" I share this example to illuminate how creating online spaces where the experiences and knowledges of ITCs can be shared can lead to an environment where ITCs bring themselves into the curriculum. Although there are no guarantees with this approach, in the case study I share, this resulted in an Indigenous-centered pedagogical environment that foregrounded physical activity, health and wellbeing as experienced and determined by those in the class, not as predetermined curricular outcomes. Intentionally following Styres' design allowed for a (re)membering of the values shared within Indigenous communities that focus on relationships, history, kinship, and community physical activity, health and wellbeing while also fostering a (re)membering that enabled pedagogical self-determination to emerge.

Conclusion

While the three Indigenizing design case exemplars are quite specific, the pedagogical principles have widespread relevance for LMS design. Indigenous knowledge frameworks guided the LMS design of the mathematics on the land as well as centering land in PHE. All the Indigenous LMS design exemplars were rooted in localization. Each course was attentive to local land, stories, languages, and traditions. Engaging members of Indigenous communities ensured that their aspirations were embedded in LMS design (Ole, 2014). Multimodalities included different interactive asynchronous activities in each Indigenizing LMS design exemplar; for instance, ITCs reported their mathwalks with images and text. They also engaged in multimodal resources with video interviews and reflections on a LMS platform. The LMS design for relationships was enhanced by synchronous learning with a consistent instructor presence to foster a sense of connectedness and build relationships.

Reflections on student experiences

The Indigenizing design exemplar for Educational Psychology and Special Education (EPSE) describes part of the curriculum development for a hybrid approach. Although consultation with community partners was a starting point for this work, the inclusion of Indigenous voices in EPSE courses brought together a course instructor, educational technology designers, Indigenous teacher candidates (ITCs), and Indigenous Knowledge Keepers. The ITCs with the mentorship of the course instructor produced several interviews with Indigenous Elders and school educators. Their multimedia interviews weaved together culture and developmental perspectives. Synchronous classroom discussions and asynchronous reflections were used to engage ITCs after listening to the interviews. Yet, multimodal resources sometimes hindered the experience of some ITCs due to limited broadband internet and/or no connectivity. The multimodal resources were a catalyst for further changes in EPSE courses. It has led to discussions among faculty members about the Indigenizing design of graduate level EPSE curriculum. Through discussions with faculty, there has been an emphasis on disrupting Western, educated, industrialized, rich, and democratic (WEIRD) resources widely used in curriculum. That is, along with the Indigenizing of teacher education curriculum, there are discussions about how to decolonize online learning spaces. We encourage other educators to consider how to disrupt taken for granted LMS practices as a way of decolonizing education. Indigenizing design for digital mathematics education offered opportunities for ITCs to experience their communities, land, and cultural practices through a study of patterns. Building relationships and drawing upon story to explore mathematics created unexpected synergies. For example, one ITC wrote that connecting math, story, and land practices opened up possibilities to "incorporate some local Tŝilhqot'in language into the story and include local Elders to come and speak about their hunting protocols." The study of mathematics moved toward being seen as connected rather than separated from cultural practices and the land. Our goal was for ITCs to experience mathematics for living differently, to (re)story relationships with mathematics that could, in turn, support ITCs in building caring relationships with their own students, grounded in the local, and connected to land. For some ITCs this is affirmed in learning of their own students' experiences, where ITCs saw students noticing and creating patterns on the land and heard students claim, "their favorite part of the lessons were the land-based math lessons!"

Within the PHE Indigenous course design, the organization, structure, and pedagogical approach foregrounded opportunities for ITCs to share their cultural awareness while weaving in their vision and values for a holistic approach to physical activity and health. By creating the context for the emergence of strong relationship development both within the course and with community and by centering Land and Indigenous knowledge and ways of being, the course designed offered ITCs opportunities to build confidence in sharing their approach to wellness and its larger connection to self-determination.

Indigenizing pedagogy in online learning is an ongoing process. As these courses are being delivered in other NITEP field centers in different regions, there will be a need to be responsive to local Indigenous worldviews, histories, land and place. This iterative process reflects a commitment to Indigenizing within learning design. The revisioning of these courses may offer new ways that empower ITCs, schools, and communities.

Final thoughts

Indigenizing learning design for online learning assists ITCs in connecting with each other, their communities, and their lands and languages. We were deliberate that ITCs experience cultural knowledge and practices in new modalities they might not otherwise experience that includes digital stories, songs, language, or visual expressions. "Many cultural practices – at one time requiring an embodied presence – adapt to this contemporary reality" (de Haan, 2018, p. 14). As with Sam, Schmeisser, and Hare (2021), our design exemplars foster the pride among ITCs that is experienced when their storied traditions are narrated through audio, video, and multimodal forms. Space, voice, and agency are given to Indigenous people when their knowledges are upheld in digital forms. While the digital space is not a replacement for the experiential pedagogies that occur in physical and material worlds, we suggest alongside Morford and Ansloos (2021) that new relationships can be formed with land through online experiences. Digital environments serve to repatriate land, languages, and traditions (Wemigwans, 2018). The possibilities for honoring Indigenous knowledges and pedagogies in the digital space point us to the need for protocols guided by Indigenous ethics (Morford & Ansloos, 2021). This learning design approach through Indigenization is not only an inclusive approach. While our approach takes up equity and social justice, an Indigenizing design approach is an assertion of Indigenous digital sovereignty.

References

Adams, G., Dobles, I., Gómez, L. H., Kurtiş, T. & Molina, L. E. (2015). Decolonizing psychological science: Introduction to the special thematic section. *Journal of Social and Political Psychology*, 3(1), 213–238. https://doi.org/10.5964/jspp.v3i1.564

Archibald, J. Q. Q. X. (2008). *Indigenous storywork: Educating the heart, mind, body, and spirit.* UBS Press.

Armstrong, J. C. (1998). Land speaking. In S. Ortiz (Ed.), *Speaking for the generations: Native writers on writing* (pp. 174–195). University of Arizona Press. https://doi.org/10.2307/j.ctv27jsm69.11

Auger, D. (2006). *Mwâkwa talks to the loon: A cree story for children.* Heritage House Pub.

Bishop, A. J. (1990). Western mathematics: the secret weapon of cultural imperialism. *Race & Class*, 32(2), 51–65. https://doi.org/10.1177/030639689003200204

Brayboy, B. M. J. & Maughan, E. (2009). Indigenous knowledges and the story of the bean. Harvard *Educational Review*, 79(1), 1–21. https://doi.org/10.17763/haer.79.1.l0u6435086352229

Cajete, G. (2004). Philosophy of native science. In A. Waters (Ed.), *American Indian Thought* (pp. 45–57). Blackwell Publishing.

Cajete, G. (2015). *Indigenous community: Rekindling the teachings of the seventh fire.* Living Justice Press.

Camp, R. V. (1998). *What's the most beautiful thing you know about horses?* Children's Book Press.

Carle, E. (1994). *The very hungry caterpillar.* Philomel Books.

Castagno, A. E. (2012). "They prepared me to be a teacher, but not a culturally responsive Navajo teacher for Navajo kids": A tribal critical race theory analysis of an Indigenous teacher preparation program. *Journal of American Indian Education*, 51(1), 3–21. http://www.jstor.org/stable/43608618

Cote-Meek, S. (2014). *Colonized classrooms: Racism, trauma and resistance in post-secondary education.* Fernwood Publishing.

Coulthard, G. (2010). Place against empire: Understanding Indigenous anti-colonialism. *Affinities: A Journal of Radical Theory, Culture, and Action*, 4(2), 79–83.

Gaertner, D. (2015). Indigenous in cyberspace: CyberPowWow, God's Lake Narrows, and the contours of online Indigenous territory. *American Indian Culture and Research Journal*, 39(4), 55–78. https://doi.org/10.17953/aicrj.39.4.gaertner

Garcia, J. & Shirley, V. (2013). Performing decolonization: Lessons learned from Indigenous youth, teachers and leaders' engagement with critical Indigenous pedagogy. *Journal of Curriculum Theorizing*, 28(2).

Grafton, E. & Melançon, J. (2020). The dynamics of decolonization and indigenization in an era of academic "reconciliation." In Sheila Cote-Meek & T. Moeke-Pickering (Eds.), *Decolonizing and indigenising education in Canada* (pp. 135–154). Canadian Scholars.

Graveline, F. J. (1998). *Circle works: Transforming eurocentric consciousness*. Fernwood Publishing.

Haan, K. de. (2018). Indigenous territory in cyberspace: Exploring the cultural and metaphysical consequences of territory beyond materiality. *The Ethnograph*, 13, 14.

Hare, J. (2021). Trickster comes to teacher education. In G. Li, J. Anderson, J. Hare & M. McTavish (Eds.), *Superdiversity and Teacher Education* (pp. 36–51). Routledge.

Kimmerer, R. W. (2013). *Braiding sweetgrass: Indigenous wisdom, scientific knowledge and the teachings of plants*. Milkweed Editions.

Kirkness, Verna. J. & Barnhardt, R. (1991). First Nations and higher education: The four r's- respect, relevance, reciprocity, responsibility. *Journal of American Indian Education*, 30(3), 1–15.

Manuelito, B. K. . (2015). Creating space for an Indigenous approach to digital storytelling: "Living breath" of survivance within an Anishinaabe community in northern Michigan [Doctor dissertation, Antioch University]. https://aura.antioch.edu/etds/212

Morford, A. C. & Ansloos, J. (2021). Indigenous sovereignty in digital territory: a qualitative study on land-based relations with #NativeTwitter. *AlterNative: An International Journal of Indigenous Peoples*, 17(2), 293–305. https://doi.org/10.1177/11771801211019097

Ole, K. B. R. (2014). E-learning principles and practices in the context of Indigenous peoples: A comparative study.

Pete, S., Schneider, B. & O'Reilly, K. (n.d.). Decolonizing our practice: Indigenizing our teaching. *First Nations Perspectives*, 5(1), 99–115.

Reedy, A. K. (2019). Rethinking online learning design to enhance the experiences of Indigenous higher education students. *Australasian Journal of Educational Technology*, 35(6), 132–149. https://doi.org/10.14742/ajet.5561

Salmón, E. (2017). No word. In G. van Horn & J. Hausdoerffer (Eds.), *Wildness: Relations of people and place* (pp. 24–32). University of Chicago Press.

Sam, J., Schmeisser, C. & Hare, J. (2021). Grease trail storytelling project: Creating Indigenous digital pathways. *KULA: Knowledge Creation, Dissemination, and Preservation Studies*, 5(1). https://doi.org/10.18357/kula.149

Simpson, L. (2014). Land as pedagogy. Nishaabeg intelligence and rebellious transformation. *Decolonization, Indigeneity, Education and Society*, 3(3), 1–25.

Simpson, L. (2017). *As we have always done: Indigenous freedom through radical resistance*. University of Minnesota Press.

Skiba, R. J., Knesting, K. & Bush, L. D. (2002). Culturally competent assessment: More than nonbiased eests. *Journal of Child and Family Studies*, 11(1), 61–78. https://doi.org/10.1023/a:1014767511894

Styres, S. (2017). *Pathways for remembering and recognizing Indigenous thought in education: Philosophies of Iethi'nihsténha Ohwentsia'kékha (Land)*. University of Toronto Press.

Surkan, J. (2018). Conversations: Elmer Ghostkeeper. The Métis Architect. https://metisarchitect.com/2018/06/04/conversations-elmer-ghostkeeper/

Thijs, A. & Akker, J. V. D. (2009). *Curriculum in development*. Netherlands Institute for Curriculum Development.

Varghese, J. & Crawford, S. S. (2021). A cultural framework for Indigenous, local, and science knowledge systems in ecology and natural resource management. *Ecological Monographs*, 91(1). https://doi.org/10.1002/ecm.1431

Wagamese, R. (2019). *One drum: Stories and ceremonies for a planet*. Douglas and McIntyre.

Wemigwans, J. (2018). *A digital bundle: Protecting and promoting Indigenous knowledge online*. University of Regina Press.

Whitinui, P., France, C. R. de & McIvor, O. (Eds.). (2018). *Promising practices in Indigenous teacher education*. Springer Singapore.

William, J., DeRose, C. & Camille, C. (2018). Coyote's food medicines. Doctors of BC. https://www.fnha.ca/WellnessSite/WellnessDocuments/Coyotes-Food-Medicines.pdf

Wilson, S. (2012). *Research is Ceremony. Indigenous research methods*. Fernwood Publishing.

Hybrid Teaching is Not a Limbo Nor a Multiverse

Lessons from a HyFlex experience. Humanistic ideas for a blended future

Victor Azuaje

Welcome to the fall semester of the pandemic academic year of 2020. Twenty-five students are enrolled in your class. (This is a Spanish language course.) On Mondays, you teach half of them face-to-face and half of them online. On Wednesdays, you reverse their roles. On the other days, you interact with all of them asynchronously through content and exercises in the learning management system.

Let's assume you're lucky. On Mondays and Wednesdays, you have a desktop and a laptop in the classroom. To teach both groups simultaneously, you open your first Zoom account on the laptop to share audio and video with the online students and log in to your second Zoom account on the desktop to join the session and share the same audio and video with the face-to-face students. All this nightmarish juggling, however, solves only half of your problems: you still have to work with your students asynchronously using the learning management system. To set up the perfect storm, add two other paradoxical disturbances. First, you've been asked to keep your students engaged in every session and modality—but most researchers claim that engaging students face-to-face has few similarities with engaging them online. Second, your students didn't choose this setting: it was imposed on them because of COVID-19.

For the same unforeseen reason, you haven't been provided with a full-fledged HyFlex classroom. There was probably none. HyFlex courses weren't the outcome of an academic hype or an institutional project. In 2019, only a few institutions of higher education were offering HyFlex as an option for their students; in 2020, COVID-19 pushed many of them to place it at the center of their offerings. Because of this abrupt move, the term became the umbrella for any combination of online and face-to-face modalities. In some cases, HyFlex wasn't even a modality: it was a technical shorthand for the blender of face-to-face and online settings where anxious administrators poured in students, classrooms, and instructors, following an unsteady recipe of shrinking optimized spaces, weekly apocalyptic moods, and updated CDC guidelines. And though "HyFlex" stands here for the experience I described in the beginning, it will also refer to that manifold of mixed modalities and pedagogical practices, and their intended or improvised amalgamation.

If it is unfair to privilege HyFlex over other hybrid modalities, it is also unfair to consider on equal footing hybrid instructors and face-to-face or online only instructors. The whiteboard on Zoom isn't the whiteboard in the classroom, yet for the hybrid instructor, the two must work in tandem. But more than a divide among instructors about technology, the pandemic revealed a divide between many instructors' assumptions and most students' expectations about courses blending physical and virtual environments. Many suddenly appointed HyFlex instructors assumed a motivational and communicational gap between online and face-to-face instruction, envisioning a hybrid modality either as a limbo equipped with 14-inch-monitors and built-in microphones or as a multiverse, as separate worlds linked by rotations in the schedule and an unmanageable influx of students' messages. On the other hand, most students expected engaging hybrid instructors and consistent learning experiences across modalities, imagining instructors as producers and hosts of seamless multimedia and multi-presence shows. However, if Stephen Colbert and Jimmy Fallon weren't successful at moving their TV personas from the studio to a room in their houses with their families as live audience, why did we expect better performances from instructors switching between a classroom in their institutions and a room in their houses with no script for engaging face-to-face and synchronous plus asynchronous online audiences simultaneously?

A Unified Approach to Face-to-Face and Online Engagement in Mixed Modalities: The Hybrid Teaching Identity

Talk show hosts had it easier in the fall of 2020: they needed to keep their charm and humor in a room without a live audience. HyFlex instructors, for their part, had two types of audiences in the same room and lost more than one tool to work with half of them. There are two facts about online teaching: you can't rely on your charisma, and humor rarely works. Online students are a tough crowd. (In the 2020 fall semester I received less than twenty LOLs, and only one student wrote, "I actually laughed out loud and choked on my food when I read this email.") Extremely charismatic HyFlex instructors could then expect to be effective only half of the time with half of the crowd—the face-to-face students. Online students? They are left to an unengaging fate.

Since charisma is unreliable and a limited resource, educational researchers have focused instead on teaching identities. For the face-to-face modality, Kristy Cooper Stein et. al (Cooper et al., 2016), Rebecka Black (Black, 2018), and Tina Gutierez-Schmich (Gutierez-Schmich, 2016) offer a thoughtful survey. For synchronous and asynchronous online settings, Jennifer Richardson and Janet Alsup summarize the findings. They also share the widespread belief that condemns HyFlex instructors to the limbo or to the multiverse: "the online teacher identity…[is] significantly and necessarily different from that of a teacher in a face-to-face class; the instructor's sense of teacher self is not an extension from that in a traditional setting, but rather it is a unique creation specific to the online context" (Richardson & Alsup, 2015, p. 144). The logical fate of Hyflex instructors is sealed. If you are a face-to-face instructor, create your teaching identity. If you are an online instructor, create your unique identity. If you are a HyFlex instructor, break yourself in two—if you're a HyFlex student, say hello to Dr. Jekyll and Mr. Hyde.

Richardson's and Alsup's account takes for granted the gap between face-to-face and online teaching. But in the HyFlex context, this assumption dooms any teaching identity effort because it undermines one of its essential goals: a teaching identity, "consists of sub identities that should somehow harmonize" (Beijaard & al. qtd. in Richardson & Alsup, p. 146). Radically separating sub-identities is not part of a teaching identity project, and HyFlex instructors should not be considered an exception. They should somehow harmonize their face-to-face and online sub-identities. In doing so, however, HyFlex instructors need to find a way to incorporate Richardson's and Alsup's and similar contributions without assuming an insurmountable gap between face-to-face and online teaching. This task has now become more pressing since, "mixed-modality models, such as hybrid learning, will remain in force even as universities look beyond COVID-19" (Stone, 2021).

For my hybrid courses, I started that project by adopting the dialogical perspective promoted by Paulo Freire and the autonomy-supportive teaching style that Johnmarshall Reeve presented in, "Autonomy-Supportive Teaching: What It is, How to Do It." Freire's reflections on dialogue laid the common ground for a unified face-to-face and online identity. Reeve provided a set of instructional behaviors compatible with Freire's dialogical attitude, with my students' need for engagement across modalities, and with my concern for their emotional health in the times of COVID-19 (Reeve, 2016).

Closing the Gap Between Face-to-Face and Online Engagement: Start by Creating Your Teaching Persona

"Know thyself" may not be part of the Socratic method, but Socrates made it a relevant concern of his inquiries. Many instructors consider the search for self-understanding an indispensable practice for the profession, but for all of them, like for Socrates two thousand years and more ago, the "thyself" remains elusive. Faced with the puzzle, educational researchers have put forth proposals whose specifics are as many as the differences in the philosophical or psychological perspectives put together by theorists. Some researchers choose to work with a set of features—beliefs, values, roles—or with a more restrictive idea, the teaching persona. Both approaches are exemplified by Nancy Coppola, Starr Hiltz, and Naomi Rotter in their article, "Becoming a Virtual Professor: Pedagogical Roles and ALN." The authors are torn in two directions—e.g., the term "teaching persona" is equated with both teaching roles and teaching style—but their insights about communication highlights our need for what the Brazilian philosopher and pedagogue Paulo Freire called the dialogical attitude (Coppola et al., 2014). Before we examine how Freire can help us, let's look at the problem that requires his counsel.

Coppola et al.'s noticed how from face-to-face to online environments instructors' roles change, but they do not disappear or become indistinguishable: they remain the common denominator. And roles, in turn, have a common denominator: "Underlying the enactment of all roles is the critical factor of communication" (Coppola et al., 2014). Overt and underlying messages, from face-to-face to online environments, define the instructors' roles or what Coppola et al. labeled "teaching style" or "teaching persona." A teaching persona will then rise or fall on the quality of communication.

There is, however, a caveat. The asynchronous instructors interviewed by Coppola et al. weren't naive about the prospects of successful communication. They listed many possibilities for falling back into silence or misunderstanding. Lack or insufficiency of gestures, eye contact, voice quality, body movement, facial expressions—twenty years ago Coppola et al. heard most nowadays complaints about online communication (Coppola et al., 2014). That's why furnishing instructors with more or better gadgets or fixing their side effects should not be a post-pandemic pedagogical priority. A better 3D image resolution and stereo sound won't improve a controlling teaching style; it will only amplify or relay more clearly its disengaging messages.

This is neither a Luddite rant nor a nostalgic chant for an utter face-to-face modality. Either one is based on a fantasy. Face-to-face was never the Eden of communication. Attitudes and nonverbal clues have always hampered instructors' overall message. If online modalities have recently shown how disconnected students could be from instructors, face-to-face have long shown how alienated instructors could be from students. The hybrid environment only compounds the communication challenges already present in the two modalities.

Reading this, hybrid instructors may feel of two minds about communication. On the one hand, communication could mark the end of both the limbo and the two separate worlds, the multiverse, as metaphors for hybrid education. Through consistent messages across modalities, hybrid instructors can build a unified teaching persona. On the other hand, communication is not a magic pill: misunderstandings happen.

Let's address first a way to end the limbo or multiverse: the teaching persona. Coppola et al.'s use of the word "persona" to conflate teaching roles and teaching styles could help us set aside a misplaced emphasis on authenticity when talking about teaching identity. As Jay Parini has observed, persona (the theatrical mask, the social façade, the self-presentation) "involves artifice": "Authenticity is, ultimately, a construction, something invented" (Parini, 2005, p. 58). A romantic educational vision has highlighted authenticity or naturalness, but a hybrid teaching identity is not borne. It takes time and training to develop a new attitude.

Creating your HyFlex Teaching Persona to Engage Students: A Freirean Approach

But what kind of attitude do a hybrid instructor need to develop? Paulo Freire suggests a starting point. "Our method, then, was to be based on dialogue, which is a horizontal relationship between persons" (Freire, 2005, p. 40). Don't be deceived by the simplicity of the statement and our familiarity with the word "dialogue." First, the English translation has cleaned the definition of its original complexity: "dialogue, [...] is a horizontal relationship between persons," cleanly says the English translation; the Portuguese says it less clearly: "É uma relação horizontal de A com B" (Freire, 1967, p. 107) [It's a horizontal relationship between A and B]. The original formula has the structuralist flair so common among Latin American theorists in the 1960s, but this isn't due to the whims of a fashion. Using the sequence A-B, Freire sets out the contrast between dialogue (horizontal) and anti-dialogue (vertical). Dialogue involves horizontal relationships; anti-dialogue, "implica numa relação vertical de A sobre B" (Freire, 1967, p. 107) ["involves a vertical relationship of A over B"]. A common reading of Freire, a common structuralist mistake, is to consider A and B as poles in absolute opposition along the axes. We could see them instead as the extremes of a continuum. This point of view opens the space to different relationships of power (vertical) and to a new (horizontal), "critical and criticism-stimulating method [where] the two 'poles' of the dialogue are thus linked by love, hope, and mutual trust" (Freire, 2005, p. 40).

Second, Freire warns us that dialogue or the creation of a dialogical attitude is the most challenging task for an instructor because, "the difficulty lies [...] in the creation of a new attitude—that of dialogue, so absent in our own upbringing and education" (Freire, 2005, p. 45). We usually complain about our personal history or lack of training; Freire is more specific: we experienced a deficiency, a shortage of dialogue in our upbringing and education. Given the countless conversations we've had with instructors, classmates, relatives, and friends through the years, the claim sounds implausible. It does not if we view dialogue from Freire's perspective: a relationship mainly based on trust, empathy, and critical attitude (Freire, 2005, p. 40). Freire identified our deepest motives to hold anti-dialogical attitudes and gave us the strongest ethical reasons to let them go. He also gave us the foremost pedagogical reason to adopt dialogue as the foundation of our teaching identities: "dialogue presents itself as an indispensable component of the process of both learning and knowing" (Freire & Macedo, 1995, p. 379).

Regarding student engagement, however, Freire's comments about the role of dialogue could be baffling:

> In order to begin to understand the meaning of a dialogical practice, we have to put aside the simplistic understanding of dialogue as a mere technique. "[...] dialogue characterizes an epistemological relationship. Thus, in this sense, dialogue is a

way of knowing and should never viewed as a mere tactic to involve students in a particular task" (Freire & Macedo, 1995, p. 379)

Freire aims here neither at a technique to engage students nor at dialogue as a practice that engages students. His target is dialogue as a "mere technique" to involve students. He marks the spot where romantic notions of empathy and trust are confused with a let-them-do, I-am-feeling-good, or I-like-my-teacher pedagogy. As epistemological relationship, a relationship about how we know and learn, dialogue could make instructors and students uncomfortable with the object of study and with each other. "I engage in dialogue—says Freire—not necessarily because I like the other person. I engage in dialogue because I recognize the social and not merely the individualistic character of the process of knowing" (Freire & Macedo, 1995, p. 379). Dialogue as a practice to engage students entails taking their perspective and acknowledging their feelings of discomfort, but a dialogical practice is not group therapy. Freire shed light on the unmistakable "difference between dialogue as a process of learning and knowing and dialogue as conversation that mechanically focuses on the individual's lived experience" (Freire & Macedo, 1995, p. 381). With this distinction, he captures our sense that an instructor must empathize with her students in challenging conversations and, at the same time, assume her authority to engage them in educational directions. In this sense, Freire suggests that educators, "should avoid falling prey to a laissez-faire practice [...]. On the contrary, a better way to proceed is to assume the authority as a teacher whose direction of education includes helping learners get involved in planning education, helping them create the critical capacity to consider and participate in the direction and dreams of education" (Freire & Macedo, 1995, p. 379).

Freire tells us here how the instructor's authority must promote the student's autonomy. Unfortunately, he's been frequently misunderstood on this point. Promoting autonomy requires direction. "I do not think that there is real education without direction. [...] To avoid reproducing the values of the power structure, the educator must always combat a laissez-faire pedagogy, no matter how progressive it may appear to be" (Freire & Macedo, 1995, p. 378). Part of misunderstanding lies on the English title of his most relevant work on the subject, Pedagogy of Freedom; the Portuguese title resonates better with our purpose: Pedagogia da Autonomia (Freire, 1996).

The difference between the subtitles is more telling. The publisher in English emphasized Ethics, Democracy, and Civic Courage. In Portuguese, Freire chose Saberes Necesários à Prática Educativa: Necessary Knowledge/Wisdom for the Educational Practice. "Saber" is knowledge, know-how, and understanding, but also wisdom. "This theme of autonomy incorporates the analysis of various types of knowledge that I find to be fundamental to educational practice," Freire says (Freire, 1998, p. 21). The pedagogy of autonomy calls for "various types of knowledge," mainly know-how and wisdom. These types of knowledge/know-how/wisdom are fundamental to educational practice. And the pedagogy of autonomy is a practice for the instructor and the students: "I will know better and more authentically what I know the more efficaciously I build up my autonomy vis-a-vis the autonomy of others" (Freire, 1998, p. 87).

Autonomy-Supportive Teaching as a Dialogical Practice: A Johnmarshall Reeve's Primer

But are there any precise steps—backed by research—to start creating this dialogical teaching persona? I suggest that we begin with the ideas and practices discussed by Johnmarshall Reeve in "Autonomy-Supportive Teaching: What It Is, How to Do It." Self-determination theory (SDT) is a macrotheory—a group of six related mini-theories—of human motivation (Ryan & Deci, 2017, p. 25). SDT's main focus has been human psychological needs for autonomy, competence, and relatedness. Autonomy is the key area of research for Reeve. In his article, he explains the concepts of Autonomy Supportive Teaching (AST) style and how instructors can, "provide students with a classroom environment and a teacher-student relationship that can support their students' need for autonomy" (Reeve, 2016, p. 130).

Reeve defines that autonomy as the "need to be the origin of one's behavior. The inner endorsement of one's thoughts (goals), feelings, and behaviors" (Reeve, 2016, p. 140). He outlines how experiments on motivating styles have shown that students who receive autonomy support from instructors benefit in significant ways. Students, "experience [...] greater classroom engagement, higher-quality learning, a preference for optimal challenge, enhanced psychological and physical well-being, and higher academic achievement" (Reeve, 2016, p. 133). But not only students benefit from a more autonomy-supportive teaching (AST) environment, instructors themselves benefit when they learn or are trained to offer more support to their students' autonomy: "they further report greater need satisfaction from teaching, greater harmonious passion for teaching, greater teaching efficacy, higher job satisfaction, greater vitality during teaching, and lesser emotional and physical exhaustion after teaching" (Reeve, 2016, p. 133). As a face-to-face, online, and HyFlex instructor, my experience in adopting the AST style reduced my emotional exhaustion during the pandemic, increased my job satisfaction, and teaching efficacy.

Let's explore some aspects of AST's principles and techniques. In doing so, I'll point out how they align with Freire's ideas about autonomy and dialogical attitude based on empathy, trust, and critical attitude. I'll also suggest how an instructor can extend them to any HyFlex context as described in the introduction.

AST's focal objective is to move an instructor from a controlling motivational style to an autonomy-supportive one. Reeve assumes that both styles are the "opposite ends of a single continuum" (Reeve, 2016, p. 132). Keep in mind that sudden transformations of motivational styles aren't typical, "for most teachers, developing the skill […] occurs over time as a two-step process in which the teacher first learns how to be less controlling and then second learns how to be more autonomy supportive" (Reeve, 2016, p. 132). In other words, don't expect to create a teaching persona in a week. You're acquiring new skills; you're training your heart.

Reeve uses a grid featuring six characteristics of an autonomy-supportive teacher:

1. Takes the Student's Perspective
2. Vitalizes Inner Motivational Resources
3. Provides Explanatory Rationales
4. Uses Non-Pressuring, Informational Language
5. Acknowledges and Accepts Negative Affect
6. Displays Patience

For each characteristic, Reeve lists two or three behaviors that support student autonomy. For example, an instructor who takes the student's perspective will a) invite, ask for, welcome, and incorporate students' input; b) provide choices and options; and c) say "You may…", "You might…" (Reeve, 2016, p. 135). Taking the student's perspective falls under what Freire called empathy. Acknowledging and accepting negative affect ("Listen Carefully, Non-Defensively, with Understanding") aligns with Freire's "know how to listen" (Freire, 1996, p. 43).

These behaviors were effective in face-to-face environments. Under the assumption that they align with Freire's dialogical attitude, they could equally permeate a combination of synchronous, asynchronous, and face-to-face modalities. In designing lessons, activities or assignments, the instructor will take the students' perspective and vitalize their inner motivational resources—curiosity, relatedness, intrinsic goals. Face-to-face, emails, posts, and video messages will use non-pressuring language. Assignments in the learning management system will provide explanatory rationales first and then instructions, resources, and rubrics. Across every media and modality, students will perceive the same autonomy-supporting teaching persona.

Final Words

I have argued that the double-identity model for instructors teaching in hybrid (face-to-face and online) modalities is unsustainable: it requires instructors to summon a particular teaching identity for each modality. (I left aside the problem of how this logic, applied rigorously, multiplies our students' identities and our pedagogical quandaries.) From that point of view, the limbo option looks very appealing. Against those two visions, I've suggested a unified approach: hybrid instructors must develop a consistent teaching persona independent of technical devices and resources. As a result, students won't see a teaching style rift between face-to-face and synchronous or asynchronous online environments. To reach that goal, I've proposed that hybrid instructors adapt Paulo Freire's ideas about the dialogical attitude and Johnmarshall Reeve's strategies to develop an autonomy-supportive teaching style.

In following this path, however, we must resist the equation of the dialogical attitude and the ending of conflict and discomfort across modalities. It would be an error to mistake the demand for dialogue with the demand for common ground. Freire read the Russian philosopher and literary critic Mikhail Bakhtin, who introduced many original reflections on the notion of dialogue. As Caryl Emerson has noticed, "read Bakhtin carefully, and you will see that nowhere does he suggest that dialogue between real people necessarily brings truth, beauty, happiness, or honesty" (Emerson, 2000, p. 152). Freire read Bakhtin carefully; he would tell us that dialogue is not a panacea, but it is the beginning of a humanistic way to a hybrid teaching identity.

References

Black, R. A. (2018). *Understanding how perceptions of identity and power influence student engagement and teaching in undergraduate art history survey courses* [Doctoral dissertation, University of Arizona].

Cooper, K. S., Kintz, T. & Miness, A. (2016). Reflectiveness, adaptivity, and support: How teacher agency promotes student engagement. *American Journal of Education*, 123(1), 109–136. https://doi.org/10.1086/688168

Coppola, N. W., Hiltz, S. R. & Rotter, N. G. (2014). Becoming a virtual Ppofessor: Pedagogical roles and asynchronous learning networks. *Journal of Management Information Systems*, 18(4), 169–189. https://doi.org/10.1080/07421222.2002.11045703

Emerson, C. (2000). *The first hundred years of Mikhail Bakhtin*. Princeton University Press.

Freire, P. (1967). *Educação como prática da liberdade*. Paz e Terra.

Freire, P. (1996). *Pedagogia da autonomia: Saberes necesários à prática educativa*. Paz e Terra

Freire, P. (1998). *Pedagogy of freedom: Ethics, democracy, and civic courage*. Rowan & Littlefield.

Freire, P. (2005). *Education for critical consciousness*. Bloomsbury Academic.

Freire, P. & Macedo, D. (1995). A dialogue: Culture, language, and race. *Harvard Educational Review*, 65(3), 377–403. https://doi.org/10.17763/haer.65.3.12g1923330p1xhj8

Gutierez-Schmich, T. (2016). *Public pedagogy and conflict pedagogy: Sites of possibility for anti-oppressive teacher education* [Doctoral dissertation, University of Oregon]. http://hdl.handle.net/1794/20490

Parini, J. (2005). *The art of teaching*. Oxford University Press.

Reeve, J. (2016). Autonomy-Supportive teaching: What it is, how to do it. In W. C. Liu, J. C. K. Wang & R. M. Ryan (Eds.), *Building autonomous learners, perspectives from research and practice using self-determination theory* (pp. 129–152). https://doi.org/10.1007/978-981-287-630-0_7

Richardson, J. C. & Alsup, J. (2015). From the classroom to the keyboard: How seven teachers created their online teacher identities. *The international review of research in open and distributed learning*, 16(1). https://doi.org/10.19173/irrodl.v16i1.1814

Ryan, R. M. & Deci, E. L. (2017). Self-Determination theory: Basic psychological needs in motivation, development, and wellness. The Guildford Press. https://doi.org/10.1521/978.14625/28806

Stone, A. (2021). Making hybrid learning happen in higher Ed. *EdTech Magazine*. http://www.edtechmagazine.com/higher/article/2021/05/making-hybrid-learning-happen-higher-ed

Building a Framework
Critical pedagogy in action

Mary Mathis Burnett

For instructional designers, *the work* can sometimes feel like a series of cleanup steps to better position the course content for learners or to ease the burden of instructors. Caught in powerful systems of grading, academic integrity, and institutional hierarchy, we can find ourselves focused on consistency and efficiency rather than supporting learning as the primary goal. As an instructional designer in higher education, I am certainly guilty of missing opportunities to advocate for students in order to meet a deadline or institutional expectation, so I started asking myself what would happen if we allow *the mess* instead? What if we transformed our thinking into imagining the possibilities of pedagogy instead of reshaping and repurposing what already is? This question underwrote my goal of improving the equity, accessibility, inclusivity, and quality of instruction for online courses by enhancing the overall approachability and encouraging the broad application of critical pedagogy by faculty and instructional designers.

Critical pedagogy works to stop the ongoing damage done within the systems already at play by empowering all participants to bring about the necessary social changes they see are needed. It causes us to examine the context of the classroom, the perspectives included, excluded, and prioritized, the assumptions active in the room, and the power and authority of the people and systems within the classroom and institution (Freire, 1985). In higher education, where professional teaching practice is commonly ignored or undervalued, it is potentially unfair to ask faculty to apply critical pedagogy without support. In the interest of fostering more empowering pedagogy and instructional design, I worked to create support in the form of the *Application Framework for Critical Pedagogy* (AFCP).

As part of my doctoral journey, I designed a research study that would both examine the process of creation for a critical pedagogy framework and to describe the framework itself as created. In this chapter, I will discuss my research by describing the background and theory, the process of creation, the framework itself, and conclude with implications to the fields of teaching and instructional design.

Background and Theory

As a concept, *pedagogy* has taken various forms over many centuries, having been used in ancient Greece to describe the nurturing of the whole person independent from the teaching of content (Smith, 2019) and defined simply as, "the art, occupation, or practice of teaching," as in the Oxford English Dictionary (2020). This leaves much room to debate, disagree, practice, theorize, and apply a personal style (Smith, 2019). My training as an instructional designer centered around content, strategies, and grades, and so did my pedagogy. Encountering critical pedagogy and actor-network theory highlighted distinct gaps in a pedagogy like mine. Tending to those gaps fundamentally changed my pedagogical perspective, so my approach now requires an acknowledgement of the whole person, their agency and full humanity, their learning goals, and their ability to interact well in the world where they will encounter similar power dynamics. This acknowledgement is a step toward expanding the traditional boundaries of higher education which can be oppressive to the natural freedoms of students while also a step toward undermining the ties of higher education to the status quo.

Critical pedagogy suggests that we cannot ignore the ways we uphold white supremacy and nationalism, heteronormativity, and the patriarchy, while actor-network theory suggests that we cannot ignore the ways we establish truth and knowledge when they disenfranchise or dismiss individuals. We owe it to students to model paths to solutions, and we must involve students in the process of their own education, even if they are uncomfortable with that. I argue that critical pedagogy and actor-network theory are able to do just that and that they work together symbiotically to accomplish it.

Paulo Freire (1985), who is widely credited with founding critical pedagogy, insisted there is no neutral stance related to the freeing of the oppressed. He wrote that absolving ourselves of the responsibility to act in the conflict between the powerful and the oppressed is not neutral but an act on the side of the powerful (Freire, 1985). Without sound pedagogy, teaching becomes transactional. Technologists become investigators. Teachers become judges and enforcers. Retention and completion become synonymous with quality. Along the way, the significance of student learning is reduced, and we end up justifying Freire's concerns, particularly when it comes to minoritized students.

By the time they enter college, students understand textbooks as authority, professors as experts in the best cases (Law, 2009) and as judges of their competence in the worst, grades as the arbiters of success or failure or self-worth (Crocker, 2002), and that they must learn the right things to say or write to prove their worthiness for the grade (Illich, 1970). We are all conditioned by a system that elevates and honors knowledge without questioning how it came to be or whether it is possible for the system to operate another way. Critical pedagogy encourages us to question the status quo, looking critically at the way things are and who has the power, as opposed to the transactional model which inspires students to passively receive information.

We continue to see damage done by the white, cis-heteropatriarchy and by unchecked dominant culture on those who are oppressed and marginalized (Shah, N.D.). We watch, in real time, the destruction of voting rights for minoritized Americans, the carceral violence perpetrated and perpetuated against Black and African Americans while white owners of legal marijuana dispensaries make record profits, the overvaluation of power, aggression, and greed while belittling empathy and foresight, and the overrepresentation of white men in academia, politics, and government while the voices of women and minoritized individuals are suppressed (Leao, 2019; Rahman, 2021). Unless consciously disrupted, this status quo and the damage done by the systems are manifested in the classroom. We can see this clearly in higher education degree granting numbers wherein, "if black and Hispanic graduates earned each degree type at the same rate as their white peers, more than 1 million more would have earned a bachelor's degree" between 2013 and 2015 (Libassi, 2018, para. 3).

The system of American higher education has become heavily commodified, seeking continuous enrollment growth and promoting graduation rates as the primary measure of success. Technology to prevent cheating and surveil students controls the market and underscores the dominant culture of grades and power over humans and learning. Sean Michael Morris (2021) discussed societal acculturation to grades and what it means for students: that they attend college to succeed rather than to learn and that success is determined by grades. The power and authority of grades can overwhelm even the most well-intended teaching approach, especially when instructors across the industry have little or no pedagogical training. Ann Beck (2005) writes that schools also participate in the perpetuation of power relationships by legitimizing knowledge and practice that serve the interests of the dominant group and that critical pedagogy offers the means to equalize the classroom environment such that students can practice active citizenship, confronting and resisting the ways textbooks and discourses sustain social inequalities and injustices.

If we are to examine and challenge the power and authority in the classroom, we must also investigate the rest of our pedagogy including who holds power in each of the relationships involved. In transactional classrooms, power is necessarily in the hands of the educator and unreachable by students. It is more than the educators' power we must be aware of, though, as entire groups of people are marginalized and oppressed by the larger dominant culture. The classroom is not, nor are students, isolated from these social powers and oppressions, and educators must either uphold or upend them. Critical pedagogy wants educators to determine for themselves the future of their pedagogy.

Similarly, actor-network theory (ANT) makes three specific arguments:

1. Knowledge is not "coherent, transcendent, generalisable, unproblematic, or inherently powerful" (Fenwick and Edwards, 2014, p. 47),

2. The object or agent is not separate from the idea of it and can be realized in multiple forms, and

3. Knowledge and truth are not inherent properties but are, instead, ascribed by a network through a process (Latour, 1988; Fountain, 1999; Law, 2009; Fenwick and Edwards, 2014; Sarauw, 2016). In higher education where content matters most, ANT would question why, how, and who? Where faculty wield power over students, ANT would want to know why, how, and who?

Where ANT focuses its attention, however, is not on the knowledge itself but on agency and power, mechanisms to authorize knowledge, distribution within the network, and its purpose and effect. Fenwick and Edwards (2016) note that higher education tends to place high importance on knowledge committees, quantifiable standards, competition, and outcomes which creates and activates different assumptions than might be found outside of higher education about what knowledge is legitimate (p. 44). We find trouble in failing to realize the difference between a material or knowledge object held in different perspectives and one that manifests differently in different contexts. This idea left me with two questions: What if we let our differences coexist? What if differences were not reconciled with what we each experience but accepted as they are?

Granovetter (1973, 1983) focused on the usefulness of weak network ties that are sometimes disregarded in research in favor of much stronger relationship ties; however, he also argued that strong ties are not actually a strength for an organization. Because of the high level of similarity implied by their strength, they tend to limit the experience and exposure of individuals (Granovetter, 1983; Kezar, 2014; Liu et al., 2017). Weak ties, on the other hand, bring diversity of thought and experience by increasing access to varied information. While strength of connection is important for network cohesion, it can undermine a network's ability to change and make progress and limit an individual's cognitive flexibility if not bolstered by weak ties (Granovetter, 1983).

Citing two ethnographic studies by Stack (1974) and Lomnitz (1977), Granovetter (1983) highlights the necessity of strong ties for poor, marginalized, and insecure populations. They have fewer alternatives in the form of weak ties, relying on their close relationships for survival in times of food and housing insecurity, while the wealthy are privileged to be able to explore their weak ties to transform their worlds without concern for their survival. As the network strengthens its ties and minimizes its weak ties, it also reduces its ability to access information born outside of the network (Granovetter, 1973; Granovetter, 1983; Kezar, 2014; Liu et al., 2017), and therefore, the status quo is reinforced.

Critical pedagogy stands in the margins demanding an end to oppression, asking to be deployed on behalf of the oppressed rather than simply asking the oppressors to change. It promotes reminding marginalized persons of the power they already have, so they are able to speak in their voices and stand for themselves. Freire called for engaged praxis rather than limiting ourselves to theoretical rhetoric, and that is where this study begins. Alongside critical pedagogy, social network theory examines all players involved in a given group, connections to each other and the outside world, and the flow of knowledge, power, and material, while actor-network theory equalizes power and examines all players, their complex social lives and experiences, and the way knowledge is accepted and power given.

Creating the Framework

As an instructional designer in a large, public, research university, I work with around 430 faculty across Watts College of Public Service and Community Solutions, many of whom teach online and face-to-face. My purpose in setting up this project was to improve the pedagogical experience of faculty and instructional designers in the college. With access to training resources and support but no requirement to participate, faculty engagement is low across the board. Gardiner (2005) wrote about this as a pervasive higher education problem.

> "Higher education continues its longstanding custom of investing little in the preparation of its teachers for their work as educators...Where faculty and staff professional development programs exist, more often than not they are weak, participation in them is voluntary, and they are given only desultory moral and financial support by senior administrators" (Gardiner, 2005, p. 12).

Courses in this college place a priority on traditional lectures, papers, and presentations and generally lack critical pedagogical application. We aimed both to challenge faculty to move beyond papers and presentations and to improve the use of critical pedagogy in online courses.

Applying tenets of critical pedagogy, actor-network theory (ANT), and social network theory (SNT), I was interested in how such a framework would be created, identifying the social, cultural, environmental, and relational factors which influence the process, and encouraging broader use of critical pedagogy. I also hoped to examine and understand the nature of power and privilege at work in the college and to create a leveling effect whereby all voices could have equal priority. It was in the active, intentional deployment of equitable practices where critical pedagogy and actor-network theory complemented each other, in two main ways: 1) ANT joined critical pedagogy to explicitly apply equal priority to the voices involved, and 2) ANT connected with social network theory to review, explore, and describe the social and relational influences involved in creating the Application Framework for Critical Pedagogy (AFCP).

I used Participatory Action Research (PAR) as the foundation for the project which asserts that mutual understanding and collaboration lay the path toward new knowledge and allow the researchers and participants to share ownership of the research project and determinations of its success or failure (Grant et al., 2008, p. 590). Influenced early by Freire's idea of conscientization, PAR came to represent the empowerment of marginalized people rather than using an outsider as savior, the idea that they are able to design action based on their own critical analysis (Coughlin and Brydon-Miller, 2014), and the aims to support their ability to do so (Grant et al., 2008). The principles of critical pedagogy and participatory action research suggest that student-, faculty-, and staff-stakeholders must be included in processes that will affect them, and actor-network theory provides the means to make those efforts successful.

The first step for us was to gather information from the community members which included students, faculty, and staff across the college. We needed to understand how familiar they were with critical pedagogy, how comfortable they were with the language and application of critical pedagogy, whether their comfort level changed depending on whether they were teaching face-to-face or online, and what they actually wanted from the classroom. Given the subjective nature of critical pedagogy and the personal nature of pedagogy in general, the priorities of the community members were at the top of the list of things any framework should provide. I used an online survey to retrieve those answers and followed the survey with interviews of those who volunteered. I was able to put together a diverse team representing each of the four schools in the college: School of Social Work, School of Criminology and Criminal Justice, School of Public Affairs, and School of Community Resources and Development.

I created a Slack workspace so those on the team could connect via online chat to focus their list of priorities so we had an agreed upon starting point. The Slack workspace was set up to neutralize many of the known power dynamics by operating in small, anonymized groups so the discussion could move forward without their knowledge of each other's gender, race, tenure status, or status as faculty, staff, or student. I'm including the discussion of Slack, despite that it did not go well for this team, because I believe anonymizing interactions at some point in the process of creating a framework like this is important to mitigate the current power and social dynamics if possible. Because the Slack discussions yielded so little information, I had the team rank the importance of characteristics of critical pedagogy that I pulled from the early interviews and survey data.

During the active creation process, the committee defined evaluation, the practice of critical pedagogy, the expectations and layout of the framework, and the way the framework would be evaluated and implemented. The committee was composed of 12 individuals, including myself, from all four schools in the college as faculty and staff. Notably, students were missing from these discussions, though two students had participated in the interviews and three others completed the survey. While the hope had been to have students involved throughout the build sessions so they could carry their own message, the addition of actor-network theory created a path for me to ensure their thoughts were represented in the build sessions after they stopped participating. Additionally, four of the faculty and staff present in the meetings also identified as students and were able to bring their experiences into the discussions, while others included feedback and information they had received from students in their courses.

Early on, it became clear that a simple visual framework or checklist would not accomplish what we had set out to create. Folks in the meetings shared the ways they work, characteristics of effective tools, and how they would or would not be able to use this framework as we discussed the language and purpose of the tool. In addition, the build team focused heavily on the priorities, outcomes, and structure of the tool and decided to employ a question-based format to encourage reviewers to decide for themselves how well the course applied critical pedagogy. This would also allow for professional growth through the process and provide scaffolded support for those faculty who need basic pedagogical training and those who might apply more advanced pedagogies. The framework needed to be somewhat self-contained, shareable, offer results on a continuum along with resources to match, require and encourage critical self-reflection on the part of the faculty, and not let them "off the hook" by allowing them to feel exempt from the equity, ethics, and social justice issues at hand.

The Application Framework for Critical Pedagogy

Considering all of the expressed needs, I built the Application Framework for Critical Pedagogy using Twine, an online browser-based system designed for interactive story-telling. The tool uses a primarily text structure, is open-source, and available to anyone with internet access. Twine includes rich support resources and is useful even for those with no technical expertise or background. This tool allowed me to build solutions quickly and to create a branched decision-making functionality where users could determine their own starting and ending points.

After an initial pass at creating the framework based on the discussion from the first build meeting, I brought it to the committee for feedback. There was general approval for the format, layout, structure, and direction, but there was much discussion about particular word choices made for certain sections of the framework. The word power, for instance, garnered some specific attention during the discussion.

> P2: Is anyone concerned with faculty looking at the word student power and frowning? I only say that because I think we need to empower, right? I don't want someone to- those that aren't invested in this process, we need to capture, and I don't

want them to get the wrong idea of what the goal is here. It isn't for the student to take power over the class. It's to empower them to feel engaged.

P1: The term I use in my class is; you take ownership of your learning. You take ownership of your project. I could definitely, [Participant 2], you're in [School of Criminology and Criminal Justice] like me, us [criminal justice] folks don't like anything that's going to- the power thing won't go well with the majority of our faculty.

Participant 9 also cautioned the group against contributing unexpectedly to the marginalization of already marginalized students.

So there's been cases where, you know, we have used peer evaluations as part of the scoring. And, you know, we will take the workgroup and have them evaluate each other. So I do agree that peer evaluations and peer feedback is critical in this shared power environment and it's a great learning tool. However, I've seen that marginalization has occurred in those situations as well, to further oppress or marginalize those that are [already] marginalized, so that...was [an] unexpected consequence. I think, I know that was my blind spot, but so how we weight that is important.

The final build session began with an overview of the final draft of the framework, displayed in Figures 1-5, for review by participants. More than anything else, this session involved tying up some of the loose ends such as deciding how we would pilot test the final framework and what adjustments to the structure or function were necessary. After the discussion of the word power in the second, I replaced it with the word autonomy to accommodate the group's feelings about the implied construct of power. Power dynamics being an essential part of any critical pedagogy review, the discussion was important to frame the concept in an appropriate way for a diverse group of faculty. Interestingly, those who expressed an initial reaction to the word power were unbothered by autonomy, but those unbothered by power reacted strongly to autonomy.

Figure 1 shows the overall structure of the framework which had become more streamlined and focused since the first draft. It includes the various pathways a reviewer might take in the process, including a sidebar with helpful information about critical pedagogy and the tool itself in addition to the review paths selected by the reviewer. The build

Toward a Critical Instructional Design

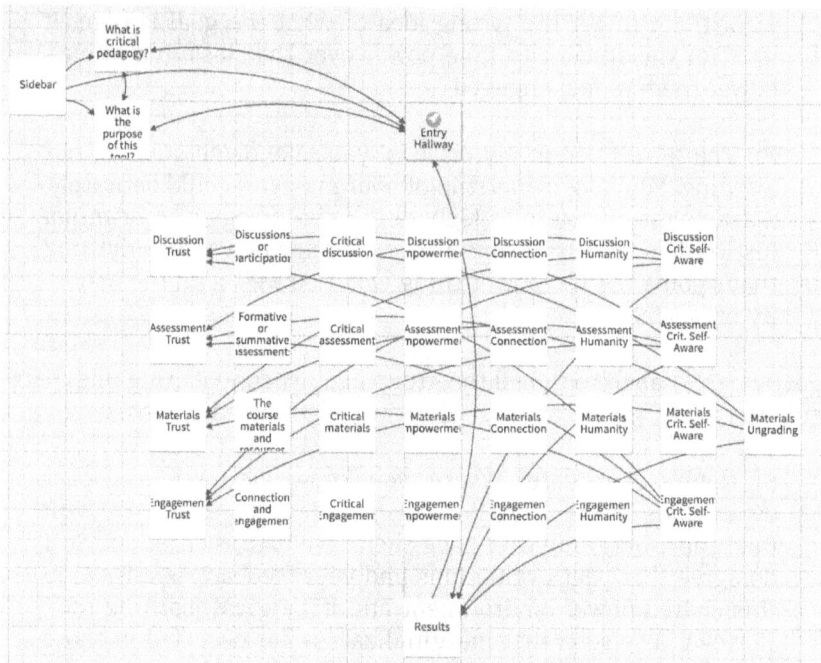

Figure 1: Screencapture of the Overview Page of the Final Draft of AFCP

team decided on four common categories in use with online courses (i.e. Discussions, Assessments, Materials, and Engagement) and seven characteristics of critical pedagogy with which to review courses (i.e. Trust Building, Critical Thinking, Empowerment, Connection, Human Experience, Critical Self-Awareness, and Ungrading [only for the Materials category]). The results page processes the decisions made by the reviewers to provide resources that may assist in improving the application of critical pedagogy for the specific course reviewed.

Figure 2 shows the initial landing page for reviewers which operates as a welcome, an introduction, and the navigational starting point for each section of the framework. From here, reviewers determine their own needs and navigate accordingly to find help, critical pedagogy resources, and sections for review.

Figure 3 shows the What Is the Purpose of this Tool? page. This page offers a brief description of the tool, its intention and design, the same critical pedagogy resources section as on the starting page, and a return to the starting page from which reviewers are able to navigate.

Figure 4 shows the What is Critical Pedagogy? page which houses a handful of curated video resources which focus on different aspects

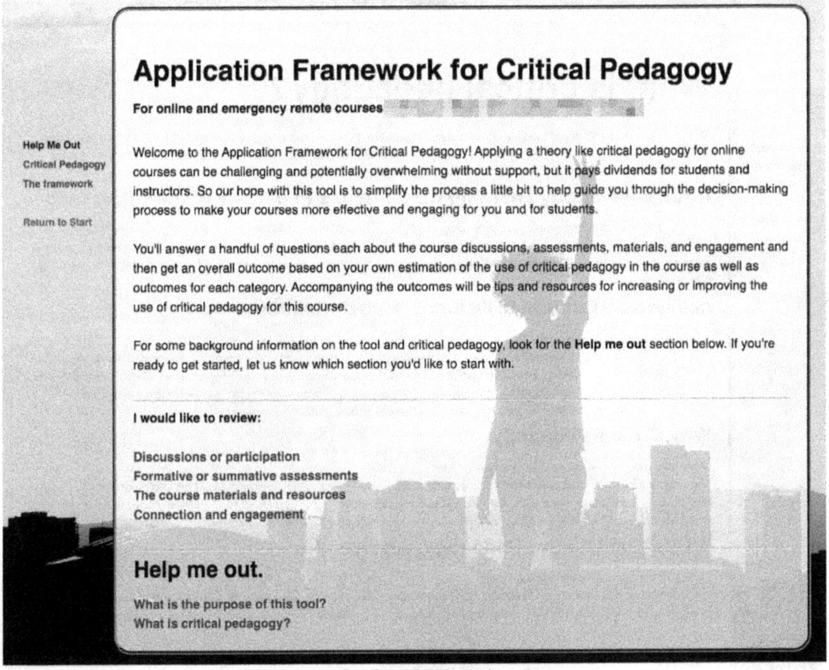

Building a Framework

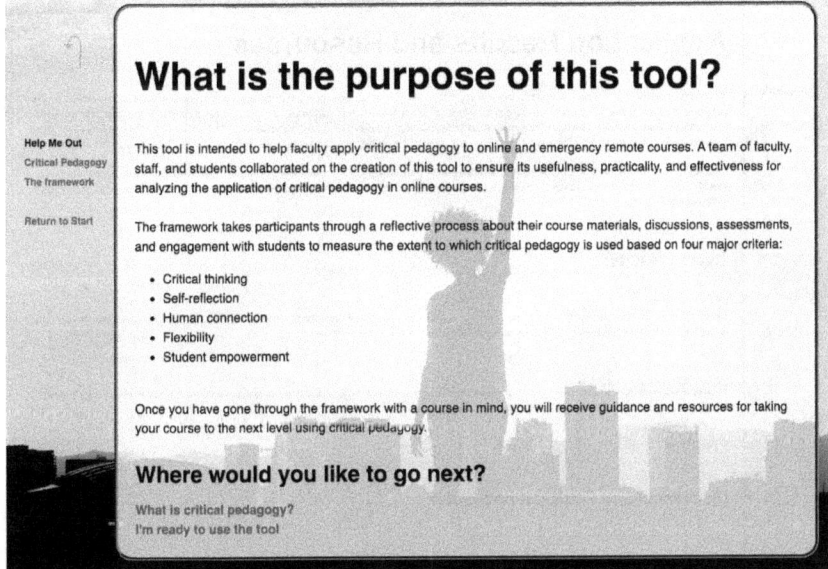

Figure 2: Screencapture of Final Draft of AFCP Starting Page

Figure 3: Screencapture of Final Draft of AFCP Purpose Page

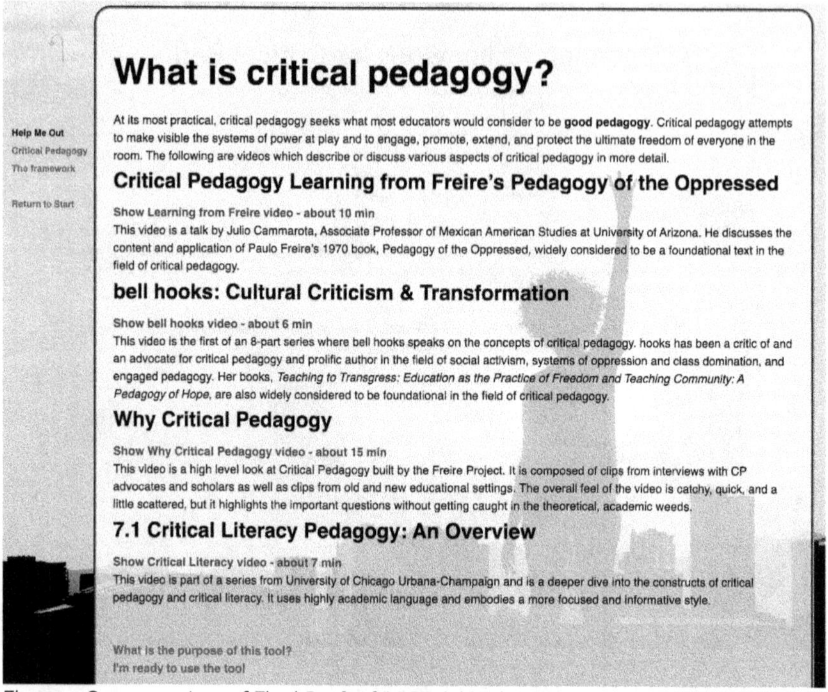

What is critical pedagogy?

Help Me Out
Critical Pedagogy
The framework

Return to Start

At its most practical, critical pedagogy seeks what most educators would consider to be **good pedagogy**. Critical pedagogy attempts to make visible the systems of power at play and to engage, promote, extend, and protect the ultimate freedom of everyone in the room. The following are videos which describe or discuss various aspects of critical pedagogy in more detail.

Critical Pedagogy Learning from Freire's Pedagogy of the Oppressed

Show Learning from Freire video - about 10 min

This video is a talk by Julio Cammarota, Associate Professor of Mexican American Studies at University of Arizona. He discusses the content and application of Paulo Freire's 1970 book, Pedagogy of the Oppressed, widely considered to be a foundational text in the field of critical pedagogy.

bell hooks: Cultural Criticism & Transformation

Show bell hooks video - about 6 min

This video is the first of an 8-part series where bell hooks speaks on the concepts of critical pedagogy. hooks has been a critic of and an advocate for critical pedagogy and prolific author in the field of social activism, systems of oppression and class domination, and engaged pedagogy. Her books, *Teaching to Transgress: Education as the Practice of Freedom* and *Teaching Community: A Pedagogy of Hope*, are also widely considered to be foundational in the field of critical pedagogy.

Why Critical Pedagogy

Show Why Critical Pedagogy video - about 15 min

This video is a high level look at Critical Pedagogy built by the Freire Project. It is composed of clips from interviews with CP advocates and scholars as well as clips from old and new educational settings. The overall feel of the video is catchy, quick, and a little scattered, but it highlights the important questions without getting caught in the theoretical, academic weeds.

7.1 Critical Literacy Pedagogy: An Overview

Show Critical Literacy video - about 7 min

This video is part of a series from University of Chicago Urbana-Champaign and is a deeper dive into the constructs of critical pedagogy and critical literacy. It uses highly academic language and embodies a more focused and informative style.

What is the purpose of this tool?
I'm ready to use the tool

Figure 4: Screencapture of Final Draft of AFCP Critical Pedagogy Overview Page

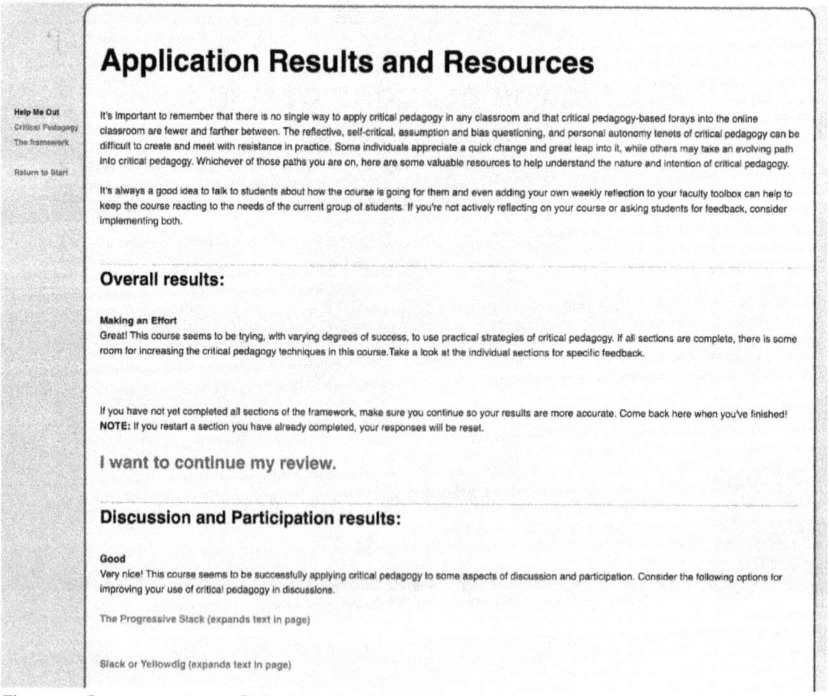

Application Results and Resources

Help Me Out
Critical Pedagogy
The framework

Return to Start

It's important to remember that there is no single way to apply critical pedagogy in any classroom and that critical pedagogy-based forays into the online classroom are fewer and farther between. The reflective, self-critical, assumption and bias questioning, and personal autonomy tenets of critical pedagogy can be difficult to create and meet with resistance in practice. Some individuals appreciate a quick change and great leap into it, while others may take an evolving path into critical pedagogy. Whichever of those paths you are on, here are some valuable resources to help understand the nature and intention of critical pedagogy.

It's always a good idea to talk to students about how the course is going for them and even adding your own weekly reflection to your faculty toolbox can help to keep the course reacting to the needs of the current group of students. If you're not actively reflecting on your course or asking students for feedback, consider implementing both.

Overall results:

Making an Effort
Great! This course seems to be trying, with varying degrees of success, to use practical strategies of critical pedagogy. If all sections are complete, there is some room for increasing the critical pedagogy techniques in this course. Take a look at the individual sections for specific feedback.

If you have not yet completed all sections of the framework, make sure you continue so your results are more accurate. Come back here when you've finished!
NOTE: If you restart a section you have already completed, your responses will be reset.

I want to continue my review.

Discussion and Participation results:

Good
Very nice! This course seems to be successfully applying critical pedagogy to some aspects of discussion and participation. Consider the following options for improving your use of critical pedagogy in discussions.

The Progressive Stack (expands text in page)

Slack or Yellowdig (expands text in page)

Figure 5: Screencapture of Final Draft of AFCP Results Page

Building a Framework

and perspectives of critical pedagogy to provide context for reviewers who are unfamiliar with the theory.

Figure 5 shows an example of the Results page, in this case having calculated only the Discussion and Participation section. Reviewers have the opportunity to decide to continue their review of the remaining sections or to explore the resources offered in the discussion results.

With any luck, the AFCP will have no final state, but in its final form for our project, the framework asks reviewers for their own assessment of a course's use of each characteristic of critical pedagogy within four categories: discussions, assessments, materials, and engagement. The AFCP does not offer scores but does use the reviewers' answers to estimate the course's success in applying critical pedagogy (i.e. Exemplary, Making an Effort, Missing the Mark, and Not Trying). Based on that estimation, it offers resources for the user to apply or adapt for their own purposes, leading with the philosophy of critical pedagogy rather than a prescriptive to-do list. It bears pointing out that the rating system here may be akin to grading, because we were working within the limitations of Twine, my experience with Twine, and dissertation time constraints. In future iterations, I hope to find a better way to match appropriate resources with reviewers' answers.

We had eight Watts College community members who did not participate in the build sessions to pilot test the framework. After using the AFCP to determine how well a course applied critical pedagogy, these participants completed an online survey to report their results and impressions. Here, I was looking for whether they agreed with the framework results and what they thought of the overall experience. Participants' reported their experience with the framework with questions about their familiarity with critical pedagogy, aptness of the framework's structure and content, likelihood of applying critical pedagogy in the future, accuracy of results, and ease of use.

This version of the Application Framework for Critical Pedagogy does meet many of the expressed needs of the committee with its flexible and self-directed function, offer of basic critical pedagogy and other recommended resources, and language encouraging users to challenge themselves, their thinking, and the status quo. The AFCP does not yet meet every need expressed by the participants; however, with the flexibility of format, it could easily have additional pedagogical resources installed, a more visually diverse interface, and the ability to be used in more discrete and nuanced ways.

Implications for Instructional Design and Pedagogy

As instructional designers, we can exercise some influence over the direction a course takes and the practices necessary to run it well, and that influence can vary by institution, program, course, and even faculty member. Part of the instructional designer's job is to ensure the instructional soundness of a course which requires an eye on the student and faculty experiences, pedagogical choices, programmatic and course-level outcomes and priorities, and technology functionality. The AFCP (https://critpedframework.com) can be a tool for instructional designers to use as part of their quality assurance process or as a teaching tool for faculty with whom they work.

One of the stated requirements for the AFCP was to promote good pedagogy more generally, so instructional designers can use the framework to proactively identify principles and practices of pedagogy that could be applied to individual courses in addition to promoting the ideals of critical pedagogy. The framework is designed as a thinking tool which can promote the type of critical thinking about courses that the courses themselves are wont to produce.

While the framework has not been distilled to simply good practice pedagogy, one of the primary goals was to allow for faculty to evolve into the use of critical pedagogy. In practice, critical pedagogy requires a certain readiness to self-reflect in a critical way, so asking a new faculty member with no pedagogical training to do this from the beginning is unfair both to the instructor and to their students. To encourage professional growth, the committee curated resources to meet faculty wherever they are in their journey. Because the framework is designed as a reflection and thinking tool, it can also be used as an opportunity for faculty to hear feedback they might otherwise have rejected.

This project connected a team of people from diverse backgrounds and sometimes opposing perspectives to distill the principles and practices of critical pedagogy into an accessible format without avoiding or reveling in its political foundation. Research for the framework also continues with an expanded audience outside the university. The AFCP is a step toward change for those faculty and instructional designers who want to participate in social changes but may not know where to begin in their own context. In this functional framework, at least, we have provided an accessible means for folks to encounter critical pedagogy, a meaningful mechanism to evaluate performance, and an opportunity to enhance learning by exposing power dynamics and encouraging self-reflection and self-advocacy. Future researchers, faculty, staff, and students are encouraged to use both the process and tool for their own contexts and purposes.

References

Beer, S. (2001). What is cybernetics? [Address]. *Kybernetes* 31(2). doi: 10.1108/03684920210417283. https://web.archive.org/web/20160426001835/http://www.nickgreen.pwp.blueyonder.co.uk/beerWhatisCybernetics.pdf

Berger, D. (2021, January 11). The multiple layers of the carceral state. *Black Perspectives*. https://www.aaihs.org/the-multiple-layers-of-the-carceral-state/

Caldwell, J. (1970, April 29). *Proceedings from American Council Fellows closing seminar: The role of higher education in social change* [keynote]. Raleigh, NC. https://soh.omeka.chass.ncsu.edu/items/show/33094

Coghlan, D., & Brydon-Miller, M. (2014). Participatory action research. *The SAGE encyclopedia of action research* (Vols. 1–2). Sage Publications Ltd. https://dx.doi.org/10.4135/9781446294406

Crocker, J. (2002). The cost of seeking self-esteem. *Journal of Social Issues*, 58(3), 597–615.

Dewey, J. (1925). *Experience and nature*. George Allen & Unwin Ltd.

Fenwick, T. & Edwards, R. (2014). Networks of knowledge, matters of learning, and criticality in higher education. *Higher Education*, 67, 35–50.

Fountain, R.M. (1999). Socio-scientific issues viz actor network theory. *Journal of Curriculum Studies*, 31(3), 339–358.

Gamoran, A. (2018, May 18). The future of higher education is social impact. *Stanford Social Innovation Review*. https://ssir.org/articles/entry/the_future_of_higher_education_is_social_impact

Gardiner, L.F. (2005). Transforming the environment for learning: A crisis of quality. In S. Chadwick-Blossey (Ed.), *To improve the academy: Resources for faculty, instructional and organizational development* (Vol.23). Anker Publishing Company. http://citeseerx.ist.psu.edu/viewdoc/download;jsessionid=C20901408F7B921A8C094ACCB7022D13?doi=10.1.1.520.680&rep=rep1&type=pdf

Ginder, S. (2014). *Enrollment in distance education courses, by state: Fall 2012.* U.S.

Department of Education, National Center for Education Statistics, Institute of Education Sciences. https://nces.ed.gov/pubsearch/pubsinfo.asp?pubid=2014023

Giroux, H.A. (2017). Critical theory and educational practice. In Darder, A., Torres, R.D., Baltadano, M.P. (eds.) *The critical pedagogy reader*, (3rd ed.). (31-55). Routledge.

Granovetter, M.S. (1973). The strength of weak ties. *American Journal of Sociology*, 78(6), 1360-1380.

Granovetter, M.S. (1983). The strength of weak ties: A network theory revisited. *Sociological Theory*, 1, 201-233.

Grant, J., Nelson, G. & Mitchell, T. (2008). Negotiating the challenges of participatory action research: relationships, power, participation, change and credibility. In Reason, P., & Bradbury, H. *The SAGE handbook of action research* (pp. 588-601). Sage Publications Ltd. https://dx.doi.org/10.4135/9781848607934

Greene, M. (2007). *Countering indifference – The role of the arts*. https://maxinegreene.org/uploads/library/countering_i.pdf

Heath, C., Hindmarsh, J., & Luff, P. (2010). *Video in qualitative research*. Sage Publications, Inc. https://dx.doi.org/10.4135/9781526435385

Hidebrand, David (2018, November 1). John Dewey. *Stanford encyclopedia of philosophy*. https://plato.stanford.edu/entries/dewey/

Illich, I. (1970). *Deschooling society* [Kindle for iPhone version]. KKien Publishing International.

Kemmis, S. (2008). Critical theory and participatory action research. In Reason, P., & Bradbury, H. *The SAGE handbook of action research* (pp. 121-138). Sage Publications Ltd. https://dx.doi.org/10.4135/9781848607934

Kezar, A. (2014). Higher education change and social networks: A review of research. *The Journal of Higher Education*, 85(1), 91-125.

Kim, J. (2019). The conversation about scaling high-quality/low-cost graduate online education. *Inside Higher Ed*. https://www.insidehighered.com/blogs/technology-and-learning/conversation-about-scaling-high-quality-low-cost-graduate-online

Latour, B. (1988). *Science in action*. Harvard University Press.

Law, J. (2009). Actor network theory and material semiotics. In B.S. Turner (ed.), *Blackwell companions to sociology: The new blackwell companion to social theory*. (pp. 141-158). Wiley-Blackwell.

Leão, G. (2019, August 20). Understanding the rage of white male supremacy: An interview with Lisa Wade, PhD. *WMC Women Under Siege.* https://womensmediacenter.com/women-under-siege/understanding-the-rage-of-white-male-supremacy-an-interview-with-lisa-wade-phd

Libassi, C.J. (2018, May 23). The neglected college race gap: Racial disparities among college completers. *American Progress.* https://www.americanprogress.org/article/neglected-college-race-gap-racial-disparities-among-college-completers/

Liu, W., Sidhu, A., Beacom, A.M., & Valente, T.W. (2017). Social network theory. In Rossler, P. (ed.) *The International Encyclopedia of Media Effects.* John Wiley & Sons, Inc. https://www.researchgate.net/publication/316250457_Social_Network_Theory

Mathis Burnett, M. (2020). *Building a framework: Critical pedagogy in action research* [Doctoral dissertation, Arizona State University]. ProQuest Dissertations Publishing.

Morris, S.M. (2021, June 9). *When we talk about grades, we are talking about people* [keynote presentation]. #RealCollege Virtual Journey, Philadelphia, PA. https://www.seanmichaelmorris.com/when-we-talk-about-grading-we-are-talking-about-people/

Narayanan, A. [random_walker]. (2019, October 11). *My university just announced that it's dumping Blackboard, and there was much rejoicing. Why is Blackboard universally reviled? There's a standard story of why "enterprise software" sucks. If you'll bear with me, I think this is best appreciated by talking about... baby clothes!* [Twitter moment]. https://threadreaderapp.com/thread/1182635589604171776.html

Pedagogy, n. (2020). *Oxford English Dictionary.* Retrieved from https://www.oed.com/viewdictionaryentry/Entry/139520

Rahmud, M. (2021, March 8). We are all victims of a patriarchal society: Some just suffer more than others. *Cordaid.* https://www.cordaid.org/en/news/we-are-all-victims-of-a-patriarchal-society/

Sarauw, L.L. (2016). Co-creating higher education reform with actor-network theory: Experiences from involving a variety of actors in the processes of knowledge creation. *Theory and Methods in Higher Education Research,* 2, 177-198.

Smith, M.K. (2019). What is pedagogy?. *The encyclopedia of pedagogy and informal education.* https://infed.org/mobi/what-is-pedagogy/

Swantz, M. (2008). Participatory action research as practice. In Reason, P., & Bradbury, H. *The SAGE handbook of action research* (pp. 31-48). Sage Publications

Ltd https://dx.doi.org/10.4135/9781848607934

U.S. Department of Education, National Center for Education Statistics (NCES). (2018). *Fast facts: Distance learning.* https://nces.ed.gov/fastfacts/display.asp?id=80

Wittgenstein, L. (1953). *Philosophical investigations.* Basil Blackwell Ltd.

On Practice: Instructional Designer

Autumm Caines

Instructional Design
Born of militarization
Systematic systems
Engineered acquisition

Welcoming of technology
(Another kind of weapon)

If it wasn't for the humanness
It would be so orderly
So refined

What even is an Instructional Designer?

Student. Teacher.
We know these archetypes
But what of this third entity
This interloper
On how we come to know

You may be told it is all wrapped up
In knowing what
Maybe knowing how
And, too often, not so much in knowing how to know

What even is "content"?

Remember; Remember; Remember
Remember till you realize the value of forgetting
Just to remember
Again

So...
As it happens
I am an Instructional Designer
Here at the threshold
All in-between

Armed with the munitions:
Outcomes, Objectives, Goals
Behaviorist
Constructionist, Constructivist, Connectivist
LMS, VLE, LTI

All the next generation
Innovation
Acronyms
And buzzwords

You should see me iterate

But something is off
In these theories and tools

When we analyze these systems
We often find
Horrible things
Unintended?
But look who benefits!

When we slow down
When we think about those left out
When we imagine something different
Than the stated objectives

When we scrutinize the tool
Be that tool the theory or technology
And use our voices
To challenge flawed systems

Are we doing the work?

I think so

Because what even is an Instructional Designer?
If not human

Quality Theater

A post-rubric approach to online course quality

Jerod Quinn

There's a phrase that has stuck with me since I first heard it a few years ago; security theater. It's looking like you are taking active approaches to create a safer environment, except those actions are not really creating a safer environment. They just have the *appearance* of security, *without actually making things more secure.* I think we hold this same theatrical, wishful thinking in our rubric-based approaches of ensuring the quality of an online course. We make a lot of noise and movement, spending hours on preparing a course for a quality review, coaching the instructor on how to pass the review, reviewing the course, fixing the course, and then re-reviewing that same course. But most of our energy is spent checking off items on a list that have very little impact on the online learner's experience. Instructional designers have become actors in our very own Quality Theater. If we take a critical look at what is a ubiquitous and often assumed process in online learning, we might build a path towards a more effective, equitable, and autonomous approach that requires far less time commitment.

Building that path is the work of critical instructional design. It's looking at the systems we have, we build, and we propel forward. Critical instructional designers prioritize collaboration, participation, social justice, agency that builds connections between learners and instructors, but also relationships between instructors and instructional designers (Morris, 2018). The critical instructional designer rebels against the standardization of one-size-fits-all processes and technologies, because when we choose practices that are rooted in equity and autonomy, in collaboration and self-determination, in the back-and-forth dialogue of critical consciousness, we can build something inclusive and powerful (Morris, 2018). A deeply common and, I would argue, deeply problematic area of instructional design in education that needs the examination of critical instructional designers is the unquestioned adoption and propagation of online course quality rubrics.

When I talk about applying a quality rubric, I'm referring to the increasingly common approach in online course quality control. The assumption is that specific design characteristics of an online course can make a better learning experience for students. A course quality rubric typically has a set of categories (like assessment) with multiple items assigned to each category. A course reviewer or multiple reviewers, sometimes a peer or sometimes an instructional designer, will search for the evidence of the specified criteria and determine if the course in question meets the standards set by the rubric. If the course meets enough of the set standards it "passes" the course quality review. These reviews focus entirely on what can be viewed in the online course content in the learning management system (LMS). They do not examine the teaching or learning practice.

To think that we can neatly separate the content and structure of a course from the teaching of a course in terms of defining what makes a quality online learning experience is an exercise in magical thinking. By ignoring the teaching, we knowingly choose efficiency of process over effectiveness of impact. There's a reason most online learning isn't in the form of a correspondence course where all the content is there for you to work through on your own with minimal, if any, instructor interaction. Interaction matters. It's not the content that predominantly makes the online learning experience work, it's the experience itself. The rubric approach to quality looks at what content is present and absent in a course, but not at what happens in the teaching of the course. While that is a very "scalable" approach to quality assurance, online learning is a slippery creature totally dependent on context and participants. It is not a widget on an assembly line that can be reproduced interchangeably with the right tooling. In this chapter, I will detail how the rubric-based quality approach to online courses is ineffective at producing engaging online courses. I will then identify what kinds of approaches might help create better online experiences that improve learner retention and motivation.

Quality Outcomes: What are We Looking to Improve?

Why go through the hassle of doing these course reviews in the first place? It's well within reason to think instructors and instructional designers can have a positive impact on the learner's experience with a pre-semester course check. The first step is to acknowledge what impact we can and can't have with a course check. We cannot instructional design our learners out of food insecurity, family caretaking responsibilities, physical and mental health challenges, or lower the outrageous costs of higher education through practices of quality assurance. These are the far more common reasons learners leave college. But we can help craft a more meaningful and supportive online classroom experience. We do this by approaching the work in a way where we can show that our efforts are worth the time investment. If a quality course review isn't improving the learner experience, then we shouldn't waste our time doing them. We need to name our outcomes.

I'm deeply tempted to point towards indicators of humanizing online learning as being our outcomes of choice. Getting individual qualitative and quantitative feedback from thousands of learners would absolutely provide rich evidence for effectiveness or lack-thereof. That would be downright glorious. But my pragmatism will not allow me to dream that big. I have never known an instructional design department to have the resources, skill sets, determination, and institutional support to be able to deploy that level of impact analysis with all the learners in all their online courses. If you can, go for it! But for the rest of us, we need a more mundane marker to use as a proxy for gauging effectiveness. I suggest learner outcomes.

When I talk about learner outcomes, I'm specifically referring to grades and drop-fail-withdraw (DFW) rates. Personally, I hate grades and the performance approaches to learning they create. But they are also pervasive and available. Anonymized grades and DFW rates can often be requested from the registrar's office without much fanfare. And like it or not, grades and GPA's are what determine if scholarships are kept, if courses need to be repeated, and even if future enrollment is allowed. Grades are the gatekeepers in determining if learners are allowed to continue their college dreams. But measuring impact from those outcomes is a tricky business.

Finding a direct line from any variable to student outcomes is a challenging task, and connecting course quality rubrics to student outcomes is no exception. Ron Legon, the Executive Director of Quality Matters™, argues that the comprehensiveness of the Quality Matters™ rubric standards makes measuring the impact of the rubric across an institution, let alone multiple institutions, a "practical impossibility" (Legon, 2015). I would argue that if measuring the impact of a quality rubric is a practical impossibility, then that is an intentional design choice in creating and maintaining that particular rubric. Course quality rubrics tend to have over 50 items, multiple standards of practice, and are divorced from an overarching learning theory. *This is critical.* Without an overarching theory of practice, there's no clear way to measure if the practice works or not. For example, you could measure motivation using a scale derived from a motivational theory, like Bandura's self-efficacy or Deci and Ryan's self-determination theory. Those theories give a marker for what counts as motivation. Without a theory for how something works, it's difficult to find evidence that it actually works. You don't know what to look for. Having such a huge number of items on a rubric also makes it overwhelming. And in the academic world, we tend to confuse being overwhelmed by information as being rigorous in knowledge. This is the "spaghetti-at-the-wall" approach to effectiveness. If we throw enough spaghetti at the wall (i.e., rubric items that are good in intention) something is bound to stick and make an impact on course quality. It's not purposeful, it's not intentional, it's not focused, but if we throw enough standards at the course something is bound to stick eventually and make the course a "quality" experience.

I will dip into some of the research on quality course reviews, but first I think there's a simpler way to make the case for why the rubric-centric approach doesn't work. I invite you to play along. Find a copy of the quality rubric you use to review online courses. If you are an instructional designer, I know you have one. Read the first checklist item out loud. Now answer this question: is that a common reason learners drop, fail, or withdraw from online courses in your context?

No?

Then why are we spending time on it as if it were?

This was the fatal crack in the dam for me. I pulled out the 48 item rubric we used at the institution where I was an instructional designer at the time, and I read through each and every item asking myself if the presence or absence of that item is something that causes learners to leave or fail online classes. I ended up identifying 4 of those 48 items that were factors in completion rates. And no, "measurable learning objectives" was not one of them. I have never heard a learner say, "I failed that online class because the instructor used the verb 'understand' in the objectives." Most of the items on that 48 item rubric were perfectly fine items. They were nominally *good things* in and of themselves. Those items were not meaningful things, not impactful things. I realized that I was spending hours of my labor per course every week with these items that don't impact the student experience in meaningful ways. I think this feeling is common no matter the particular rubric. Quality control rubrics take many shapes and forms in academia, but that approach originated with one very well-known rubric.

Quality Evidence: If it Works, We Should Be Able to Tell it Works

I've been working in online learning in higher education since 2007. In these early days of online courses the learning management systems (LMS) were often little more than a place to put folders full of files. There were few constraints and few enforced organizational structures. In practice this meant for each course everything was always different everywhere. One course I worked on in an early iteration of Blackboard's™ LMS had a system of nested folders fourteen levels deep as the organizational structure. The instructor was meeting with me because they noticed their learners were rarely turning in assignments on time if at all! I got to be the one to let them know that it was because their fourteen-level organization approach made absolutely no sense. It was in these first days of online learning when course quality rubrics started to emerge, most notably the Quality Matters™ rubric from Maryland Online in 2003 (Quality Matters, n.d.). Quality Matters™, and the many imitators that came after, created these rubrics in an effort to provide a system to "ensure course quality" for learners across the institution (Quality Matters, n.d.). As LMSs have become more constricting to what an online course can look like, the need for a checklist of present and absent items in an online course seems to be less pertinent because so much structure is dictated by the LMS. Much of the syllabus-related items and policies can be automated by the LMS administrator.

My purpose is not to dedicate this chapter to directly critiquing Quality Matters™ (QM™), but it would be impossible for me to talk about online course quality without speaking directly about them. They have been the most successful group at pushing the conversation about online course quality around the world, while also creating a market for solutions to the challenges of quality. Every other major quality rubric out there stood on the shoulders of the QM™ process and the QM™ content to create their version of a quality rubric. Countless universities have "home-grown" versions of quality rubrics based on QM™, and countless more do unofficial reviews with bootlegged QM™ handbooks. Their influence in these practices runs deep, well beyond the official Quality Matters™ certified courses. I have to talk about Quality Matters™ because so many of our unchallenged assumptions about quality online experiences come from their legacy.

There are many online course quality rubrics, most of which follow a similar construction as the Quality Matters™ rubric. Baldwin, Ching, and Hsu looked at the six most prevalent online course quality rubrics publicly available to compare their features, emphasis, and course quality criteria (Baldwin et al., 2018). They found that the rubrics had an average of 59 items, with twelve common overarching standards. These standards are:

1. Objectives are available
2. Navigation is intuitive
3. Technology is used to promote learner engagement-facilitate learning
4. Student-to-student interaction is supported
5. Communication and activities are used to build community
6. Instructor contact information is stated
7. Expectations regarding quality of communication-participation are provided
8. Assessment rubrics for graded assignments are provided
9. Assessments align with objectives
10. Links to institutional services are provided
11. Course has accommodations for disabilities
12. Course policies are stated for behavior expectations.

The application of a rubric has a similar process too. A single or multiple reviewer(s) checks for compliance to the standards, feedback is offered at least on missed standards, and the instructor is given the opportunity to revise their online course based on that feedback in order to meet the standards.

Jaggars and Xu worked to isolate which, if any, standards are directly connected to student outcomes (2016). After comparing multiple course quality rubrics they name four general categories of quality from all the collected rubrics: (1) organization and presentation, (2) learning objectives and assessments, (3) interpersonal interaction, and (4) use of technology (Jaggars & Xu, 2016). Using these categories, they examined 23 online courses and found that only one category, the quality of interpersonal interaction within a course, positively and significantly connects to student grades. They go on to explain that interaction is most impactful with frequent and effective student-instructor interactions. Despite the large collection of standards and items on quality rubrics, it seems that proportionally few of them are directly connected to improving student outcomes in online classes. In fact, they conclude by suggesting the creation of rubrics that move beyond checking for the mere presence of an item or standard and instead note how they are being used in the learning environment. They also encourage practitioners to validate each component of the new rubrics against student outcomes.

Validity and reliability are two foundational terms used in statistics and measurement conversations. Validity means your measurement tool (like a ruler) actually measures the thing it was designed to measure (like the length of a pencil). This is a crucial concept for measuring things that are hard to measure, like learning or depression. If the depression questionnaire can't help a doctor determine if someone is showing symptoms of depression then it doesn't work, it does not have *validity* in the doctor-patient context. Reliability means your measurement tool works properly over and over again with multiple populations. If we are going to claim our rubrics as measures of course quality, then they need to be valid and reliable tools.

Noting there is very little public research available on the reliability and validity of online course quality rubrics, Yuan and Recker completed one of the more comprehensive studies on validity and reliability (2015). They looked at the validity and reliability of their university's course quality rubric, which was adapted from the Quality Matters™ rubric, and found that one-fourth of the total items were problematic and had to be eliminated. From the remaining items they identified nine factors, explaining 73% of the total variance, with "learning activities & materials" explaining the highest amount of total variance in course quality (Yuan & Recker, 2015). Explaining variance is a good thing; it means something important is happening in that area and we should pay attention. Their conclusions state that only rubric items related to learner engagement and interaction have a significant and positive effect on online interactions, while only student-content interaction significantly and positively influence course passing rates.

There's more research out there, but the main takeaways I have found are:

- Many rubric items will not survive statistical validation (Yuan & Recker, 2015)
- Course quality rubrics average 59 items over twelve common themes (Baldwin et al., 2018)

- Learners are not in agreement with rubric publishers about which standards are most important in online courses (Ralston-Berg, 2014)
- In practice, the multiple reviewer process doesn't provide useful specific examples of aligning the item in question with the standard (Schwegler & Altman, 2015)
- Faculty resist course quality rubrics mainly because of the overwhelming time commitment involved in the process (Gregory et al., 2020; McGahan et al., 2015; Roehrs et al., 2013)
- Quality course reviews are often done in tandem with faculty development programs, so separating the effects of the programs versus the effects of the rubric is challenging if not impossible (examples of overlapping programs and their effects Harkness, 2015; Swan et al., 2012)
- Few standards connect to outcomes, mainly the standards that relate to faculty-student interactions are traceable to improving student outcomes (Jaggars & Xu, 2016)

I realize there are more reasons to use these rubrics than improving student outcomes. But honestly, I don't care about those reasons. I want to keep students enrolled and on scholarship, which means outcomes. I still believe a degree opens doors in life that are unavailable otherwise. It changes people's futures, family's futures, and the community's futures. That's why outcomes are my focus. I also realize that if the course review process is too suffocating for instructors, then we have lost the opportunity to improve outcomes for their students. We don't just need a better course quality tool, we need a better approach to course reviews. We need an approach that feels empowering as opposed to punitive.

Quality Feels: Power and Authority for the Precarious Worker

If you want to pick a fight with instructional designers, just throw shade at their quality rubrics. When I started questioning these rubrics the reactions from my colleagues were fast and furious. I was caught completely off-guard. Johanna Inman addresses this in her comments on using evidence-based research when talking with instructors about their courses. She says, "it is important to address the feelings about the evidence as much as it is to share the evidence" (Inman, 2021, p.

71). So when I push against a process of quality reviews, processes that bring feelings of clarity and authority in the work we instructional designers do, I'm also pushing against something that feels safe. I think it's challenging that feeling of safety that gets such visceral reactions from my colleagues. There is a large group of instructional designers who will defend quality rubrics with their lives. (You can find them trolling me on Twitter right now.) Instructional designers roll their eyes at me when I roll my eyes at learning objective measurability. But when I publicly question the power of the rubric, I get such a disproportionate reaction of anger as if these instructional designers gave birth to the quality control rubrics, and I'm insulting their babies. It's curious, but I think I understand why this topic stirs such forceful feelings.

I have a theory about the relationship between course quality rubrics and instructional designers. I don't have any "empirical evidence" about it, just observations and ideas. Instructional designers have expertise and experience in their profession. However, we are often working in a culture that puts us (and other staff) in a power dynamic where we are treated as subservient to the faculty (or the SME). That context implies that our expertise is without merit. That wears on you. It can make you second guess your professional (and personal) value. Instructional design is also work that takes broad expertise and some pretty good social negotiation skills to really feel like you know what you are doing. Every person you work with is different, and every context you work in is different. It is often nebulous work. That's actually what I enjoy about the job, it's always changing so it's always challenging. If you are new to the work, or something has shaken your confidence in your skill, it's easy to shame-spiral into a feeling of incompetence. Enter the quality rubric.

You know what feels authoritative? A rubric of 50-ish items, all annotated in annoying yet ambiguous detail with the word "QUALITY" in the title. Preferably in all caps. If you are the keeper of that powerful rubric, you become the expert. If you work with one for even a short time, you can get to know the standards and the required items thoroughly. The nebulous work of instructional design becomes clear in the light of rubric-based compliance mandates. Enforcing a course rubric is work that is anything but ambiguous. When your work is checking off boxes, you never really feel like an imposter. It's just box checking, after all.

When your work is checking boxes, it's really easy to quantify your value to the university administrators, few of whom are interested in the nuances of instructional design consultations. Telling the people who determine your employment status that you completed 25 quality course reviews this month and have "certified" those courses sounds concrete and valuable. Few people are asking if that work actually made a difference with the learners, but numbers and certifications are easy to support. Entire instructional design departments are anchored in these course reviews. But box checking isn't instructional design, at best it's compliance monitoring. When instructional designers push against my criticisms with such venom, it's partly because by pushing against the core work of their departments, I'm questioning their value as professionals.

Enforcing a rubric is safe. It's a lot of effort, but it doesn't demand much risk. Even when things get difficult with the instructor it's easy to deflect blame to the rubric itself or the administration that prescribed it. But it also positions the instructional designer as the gatekeeper of quality. For a staff member who is typically treated as an inferior to the great and powerful tenured faculty, it can taste pretty delicious to be in a position of authority for a change. These rubrics and their associated processes can bring a sense of value and clarity to instructional designers when so much of our work is ambiguous and often dismissed. These rubrics are safe. They are authoritative. They are powerful. They are really easy to explain to administrators who decide if your department gets to keep its funding. When the clarity and security of your career is questioned, you can get really mad, really fast.

But if we can move away from the work of compliance, we can lean into the work of instructional design. It's there we can put our energies into the things that matter most to our learners in online courses. We leverage our professional skills to help create an environment that better connects, empowers, and includes both the learners and the instructors.

Quality Focus: What Matters Most?

Earlier I noted how the rubric items that influence faculty-student connections are the ones most likely connected to outcomes. I also criticized the "spaghetti-at-the-wall" approach as being both ineffective and overwhelming to instructors. To move forward, we need to reject the 59 items on a rubric approach in favor of an approach that focuses on a smaller set of the most impactful categories on online learning: faculty-student connections, inclusion, and clarity of structure. I use the term categories deliberately because there are many practices in each of these categories that could be impactful given a specific context and population. By utilizing an approach that asks instructors to commit to categories of practice instead of prescribing universal, one-size-fits-all actions, we offer instructors and instructional designers the autonomy to leverage their professional experience in selecting (and changing as needed) the specific practices that might be the most interesting and the most effective in their own unique courses.

Faculty-student connections

The largest impact instructional designers can have on the learner experience in online classes is through creating an environment with deep and wide faculty-student connections. "Decades of research demonstrate that peer-to-peer, student-faculty, and student-staff relationships are the foundation of learning, belonging, and achieving in college" (Felten & Lambert, 2020, p. 5). Faculty-student connections tend to directly improve cognitive skill development while also directly encouraging classroom engagement (Kim & Lundberg, 2016). These same interactions also seem to be the most significant factor for both first-generation learners and non-white learners regarding positive outcomes (Felten & Lambert, 2020, p. 83). Not all interactions are positive ones, but overall, knowing that the instructors care for the learners and are actively working for learner success is the major factor in learner outcomes.

Instructional designers can assist with these connections by helping instructors create online courses full of opportunities for connection and communication. Instructional designers can equip instructors to host conversations outside the LMS using text and messaging tools like Slack, Discord, and GroupMe. But the purpose isn't the tools (just to be clear), the purpose is to help create discussion and conversation that goes beyond "post once, reply twice" to feel more like regular, dare I say *actual*, conversations. I worked with an instructor on a fluid mechanics engineering course and we created an activity called "Fluid Mechanics in the Wild." The idea was to take a selfie video where you saw fluid mechanics happening in your daily life, and then explain as best you could the principles at work. The instructor would give weekly examples of this and ask the learners to try this practice themselves. This could be the steam pipes in the engineering building or the large margarita with an upside-down bottled beer in it. These were shared and would provoke conversation (and connection) about the everyday relevance of their classwork. These connection practices don't have to be new practices, they can just as easily be a refocusing old practices: instructors participating in discussion forums instead of merely observing them, modeling discipline-specific reading practices through social annotation as opposed to just assigning journal articles, or trying contract or negotiated grading practices (think ungrading) instead of one-way evaluation. Again, we are not talking about prescribing a single required item or practice but drawing from a broad category of approaches that can create connection while offering instructor autonomy. Instructional designers can use their professional experience to help guide instructors towards more meaningful connections that build both relatedness and competence in their unique population of learners.

Inclusion

Knowing and feeling like you belong in a space is a powerful motivator. When I talk about designing for inclusion, I am referring to practices like culturally responsive teaching, accessibility, universal design for learning, trauma-informed pedagogy, and designing for neurodiverse learners. I'm casting a wide net for sure here, but I would argue that any steps that make the online classroom more inclusive are positive steps. Inclusion in these terms also means establishing a sense of belonging, which has been demonstrated as a factor in keeping learners enrolled (Tinto, 2005). That sense of belonging helps retain learners in college programs, from women in STEM (Banchefsky et al., 2019) to diverse and underrepresented groups in higher education (Thomas, 2015). Instructional designers can influence online courses through inclusive design practices. Amy Collier describes inclusive design as a practice that, "goes beyond accessibility, though accessibility is considered within inclusive design. Inclusive design celebrates difference and focuses on designs that allow for diversity to thrive. In higher education, this means asking ourselves, 'Who has been served, supported, or allowed to thrive by our educational designs and who has not?' And, 'How might we design for inclusion of more students' "(Collier, 2020)? While instructional designers can begin with content accessibility and perhaps even "Decolonizing the Syllabus" (DeChavez, 2018), depending on the experience of the instructor and the population of learners, we can also help guide instructors in creating approaches like land-based pedagogies (Sam et al., 2021) or legacy assessments, in which projects engage with the community and the environment and take the purpose of learning beyond the individual, toward the betterment of the community (Chavez & Longerbeam, 2016). Again, this isn't a checklist of inclusive practices to be included in a course without consideration of context. This is a wide category of options that build a sense of relatedness in the learners while instilling autonomy about the course material by being able to learn in more culturally responsive ways.

Clarity of structure

In my initial investigations of what are the most impactful factors in online learning, I tried my best to stay away from specifics on content and structure. The rubric-centric approach is entirely focused on content and structure and my hesitation was that I was trading one list of prescriptive rules for a different list of prescriptive rules. But I couldn't get away from it. Clarity of structure matters. Not having to navigate fourteen levels of nested folders on Blackboard™ matters to the learning experience. Not having to play, "read my mind" games with the instructor matters. When I talk about clarity, what I'm getting at is structure that allows autonomy while empowering learners (and instructors) to feel competent in engaging in the tasks at hand. But there's not one structure to rule them all, one template that will open access to all learners. "One right way" is the kind of thinking that replicates white supremacy principles in the workplace and in education (Okun, 2021). Course quality rubrics are rooted in the practice of "sameness" of every course in every space. The first step of critical instructional design is simply asking, "why?" Why is sameness mission critical? Do we ask on-campus instructors to conform to the practice of sameness? Do we tell on-campus algebra instructors to script their courses exactly like the anthropology or the astronomy courses? Of course not! That's ludicrous! We expect, and even encourage, on-campus instructors to shape their courses with their professional experience, in accordance with expectations of the discipline, and in conversation with their learners. So why do we prescribe sameness for those courses when they move online? This is a side effect of an earlier issue: thinking we can separate the design of the course from the teaching of the course. Templates make processes scalable, *not learning*. Learning is contextual, and we can shape that context to empower and encourage learner participation. Clarity of structure simply means that the learning process should not be an unsolvable mystery to the learners. It does not mean that everything is always the same. Clarity happens within the chosen structure, not from requiring the same structure.

Another way of thinking about structure would be in terms of making implicit expectations explicit instructions. For example, instructional designers can share the Transparency in Teaching and Learning (TiLT) framework for assessments with instructors. This is an approach that doesn't require instructors to change their assignments, but provides a structure to explain the expectations. TiLT asks instructors to describe the assignment in three sections: purpose, task, and criteria for evaluation (Winkelmes et al., 2019). This framework with its relatively simple implementation tends to improve learners' confidence, belonging, and skill development, but shows even more benefit for learners typically underserved by higher education institutions (Winkelmes et al., 2016). Instructional designers can use their professional experience to help guide instructors towards content structures and greater clarity that empower their unique population to navigate the experience with competence and autonomy.

Overarching theory

Earlier in this chapter I claimed, "without an overarching theory of practice, there's no clear way to measure if the practice works or not." So what would be the overarching theory that provides a framework for measurement of effectiveness for the collection of faculty-student connections, inclusion, and clarity of structure? I think there are multiple frameworks that could potentially work, but for me as a quantitative-minded educational psychology researcher interested in motivation and learning, the immediate connection is the self-determination theory of motivation. Self-determination theory (often abbreviated as SDT) states that people have a motivational need for competence, autonomy, and relatedness, and these three needs can guide a person's behavior (Schunk et al., 2014). Individuals need to feel competent when interacting with others and when engaging with various tasks and activities. The need for autonomy refers to the need for a person to feel a sense of control or agency in their interactions with the environment. A sense of relatedness would feel like belonging to a group of people, being included and connected to others (Schunk et al., 2014). I see pairs of SDT items working together in each of the high-impact categories.

- **Faculty-student connection = relatedness + competence**. Relatedness because of the personal connection with the instructor, but also competence because that connection can lead to clarity of understanding of the course material.
- **Inclusion = autonomy + relatedness**. Inclusion can bring autonomy when we look at factors like accessibility, universal design for learning, and neurodiverse instructional approaches. It also leads to feelings of relatedness through culturally responsive instruction and creating a learning environment where learners can bring as much

of themselves as they want to the experience.

- **Clarity of structure = competence + autonomy.** Clear instructions and transparent expectations let the learners know without ambiguity what they are responsible for. A navigable structure also creates marked paths for completing the course and engaging with the materials.

Self-determination theory has been used in educational research in multiple contexts, including online learning, to measure engagement and effectiveness of approaches (Hsu et al., 2019; Niemiec & Ryan, 2009; Wang et al., 2019). While SDT is my choice of theoretical framework, it's far from the only candidate. The point is in order to claim effectiveness, quality, or impact, we need a way to name and measure those factors that contribute to being effective, quality, and impactful. Otherwise, we risk spending our time and energy putting faith in the marketing promises of companies trying to sell us their products.

I realize many of my colleagues will not hesitate to express their disgust at my desire for quantification and measurement of a new course quality process. I absolutely understand that aversion. Statistics and measurement have been wielded as a weapon of institutional compliance (and worse things) since they were created. Statistics, despite its rooting in seemingly objective numbers, is a deeply subjective approach just like every other approach we humans use to make sense of our world. Statistics, however, is also a very useful approach when we are trying to make decisions that will impact the time and energy of thousands of people. It generalizes practices into the *most* effective for the *most* people. I can't in good conscience ask the 3000 online instructors at my university to devote their time, energy, and often unpaid labor to a course quality process that merely sounds like it should work. I can't ask instructional designers to leave the main stage of the Quality Theater behind just to walk a block over to participate in the Off-Broadway Quality Theater. When we implement a system of practice that impacts thousands of educators, we need evidence that the practice makes a positive difference with learners. These online course quality practices affect thousands of educators because so many universities require some flavor of a quality check out of the fear of being out of compliance with the federal regulations.

What about federal requirements?

If you're in the United States, there are federal requirements for online courses and programs that keep those courses connected to federal funding. Back in the dark ages of online classes (2008) the federal government stipulated that in order to meet the requirements for accreditation, online classes needed to have "regular and substantive interaction" between the learners and the instructor. It took an additional decade to define what that meant.

Substantive interaction is the push to make the online learning experience connected and engaging. It requires at least two of the following: "(i) Providing direct instruction; (ii) Assessing or providing feedback on a student's coursework; (iii) Providing information or responding to questions about the content of a course or competency; (iv) Facilitating a group discussion regarding the content of a course or competency; or (v) Other instructional activities approved by the institution's or program's accrediting agency" (Code of Federal Regulations Title 34, 2022).

The definition of regular interaction is "(i) Providing the opportunity for substantive interactions with the student on a predictable and scheduled basis commensurate with the length of time and the amount of content in the course or competency; and (ii) Monitoring the student's academic engagement and success and ensuring that an instructor is responsible for promptly and proactively engaging in substantive interaction with the student when needed on the basis of such monitoring, or upon request by the Student" (Code of Federal Regulations Title 34, 2022).

Let's be honest here: I can't imagine a lower bar for regular and substantive interaction. Instructors have to give feedback on assessments *and* answer questions? Preposterous! The point of these regulations is that the Department of Education doesn't want universities to sell correspondence courses to unsuspecting learners under the banner of online courses. You don't need an extensive rubric and a drawn-out peer review process to meet the low-bar requirements of regular and substantive interaction set by the federal government. However, this is the point where I see rubric creators leveraging the administrator's fear of losing federal funds to promote their own products and services. In a recent announcement from the Online Learning Consortium about their OSCQR quality rubric, they explicitly fuel this fear by saying, "Institutions are seeking assistance in successfully navigating the new RSI [regular and substantive interaction] regulation and risk losing access to student financial aid if the institution is audited and found to be out of compliance by the DoE [Department of Education] Office of Inspector General, or as part of a periodic Departmental financial aid program review" (Chmura, 2022). Few things prompt university administrators into action as quickly as the threat of losing money. But the federal requirements are so minuscule that the quality rubric process is a monumentally over-engineered approach to what in practice is a very low expectation.

Quality Conclusion: New Ideas for a Post-Rubric Approach

There are groups out there that also recognize and are pushing against the tensions of the quality rubric approach. Peralta has moved beyond content and created a rubric for improving online equity (Peralta District, 2020). *Hybrid Pedagogy* published a piece last year from Martha Burtis and Jesse Stommel that explained their tensions with implementing an unchallenged and unquestioned rubric approach while offering a list of course design considerations for instructional designers to leverage instead (Burtis & Stommel, 2021). While I've detailed some of the research regarding the effectiveness of quality rubric processes, I also have years of experience enforcing these rubrics. I am challenging this rubric-centric approach because I have been actively involved in these reviews for much of my professional career. I have been a Quality Matters™ certified peer reviewer and been a reviewer on many Quality Matters™ course reviews. My previous institution created a customized version of the QM™ rubric, and I was a reviewer for many courses using that rubric as well. All said, I've been part of approximately 200+ online course reviews over the past 6 years. These are the primary tensions I've felt in my experience with the rubric-centric approach.

- The experience almost always feels punitive for the instructor. No matter how experienced or how skilled of an educator they are, "failing" a review feels like an attack.
- No one passes a review without coaching which requires even more time committed to the process. Even a great online educator will not "naturally" build in the 40-60 items the rubric requires because the rubric is looking for very specific, often unintuitive, things.
- Our reviews averaged around 15 hours of work each between the reviewers (not including the revision work from the instructor or pre-review coaching). This same time commitment is echoed in other groups (Gregory et al., 2020).
- There's surprisingly little published evidence that course

checks using common rubrics make a measurable impact on student outcomes or experiences. That's a lot of time and energy invested in something with fuzzy impacts.

- The prescriptive nature of the rubrics along with their processes convey, sometimes subtly and sometimes overtly, that there is "one right way" to teach or design a course. "One way" approaches are not inclusion-oriented practices.

- It universally felt like a hoop to jump through and not a genuine path to make teaching more enjoyable and learning more engaging.

- It dismisses the years of online teaching experience many faculty bring to these conversations by implying it's these items that make a course "quality" and not their skill and dedication.

The whole idea of a 50-point quality checklist with multiple reviewers that deploys a drawn-out song-and-dance review process is an approach that will never prioritize collaboration, relationship, social justice, agency, or any other values of critical instructional design. By doing these rubric-centric reviews in this manner I think we are investing in something that will never bring value on the scale of the costs. While along the way we are stealing the autonomy, competence, and relationships from our instructors in the name of scalability and having a job description we can easily quantify to administrators.

I think a better, measurable, and more effective path to online course quality lies in a process that respects the experience of the instructors, focuses on high-impact items, approaches quality in a strengths-based framework, and empowers instructors with the support and autonomy to improve the learning experience in their online courses in ways that are uniquely appropriate to their learners, disciplines, and teaching approaches. I think this would need to be a reflective and possibly co-designed approach that respects instructor experience as opposed to templated task lists. We have mounds of research to point us toward the items that make known impacts on the online learning experience. We also have a community of human-centered instructional designers with deep skill sets and even deeper passion. We have plenty of barriers to break through, but the time is now for us to create a more effective, equitable, and empowering approach to online course quality. That starts with stepping down from the stage and leaving the Quality Theater behind.

Quality References

Baldwin, S., Ching, Y.-H. & Hsu, Y.-C. (2018). Online Course Design in Higher Education: A Review of National and Statewide Evaluation Instruments. *TechTrends*, 62(1), 46–57. https://doi.org/10.1007/s11528-017-0215-z

Banchefsky, S., Lewis, K. L. & Ito, T. A. (2019). The Role of Social and Ability Belonging in Men's and Women's pSTEM Persistence. *Frontiers in Psychology*, 10, 2386. https://doi.org/10.3389/fpsyg.2019.02386

Burtis, M. & Stommel, J. (2021). The Cult of Quality Matters. *Hybrid Pedagogy*. https://hybridpedagogy.org/the-cult-of-quality-matters/

Chavez, A. F. & Longerbeam, S. D. (2016). Teaching across cultural strengths: A guide to balancing integrated and individuated cultural frameworks in college teaching. Stylus Publishing.

Chmura, M. (2022). *OLC and SUNY Online update course quality rubric based on new federal requirements for distance education.* Online Learning Consortium. https://onlinelearningconsortium.org/news_item/olc-and-suny-online-update-course-quality-rubric-based-on-new-federal-requirements-for-distance-education/

Code of Federal Regulations Title 34, § 600.2 Definitions (2022). https://www.ecfr.gov/current/title-34/subtitle-B/chapter-VI/part-600

Collier, A. (2020). Inclusive design and design Justice: Strategies to shape our classes and communities. *Educause Review*. https://er.educause.edu/articles/2020/10/inclusive-design-and-design-justice-strategies-to-shape-our-classes-and-communities

DeChavez, Y. (2018). It's time to decolonize that syllabus. *Los Angeles Times*. https://www.latimes.com/books/la-et-jc-decolonize-syllabus-20181008-story.html

Felten, P. & Lambert, L. M. (2020). *Relationship-Rich Education*. Johns Hopkins University Press. https://doi.org/10.1353/book.78561

Gregory, R. L., Rockinson-Szapkiw, A. J. & Cook, V. S. (2020). Community college faculty perceptions of the Quality Matters™ Rubric. *Online Learning*, 24(2). https://doi.org/10.24059/olj.v24i2.2052

Harkness, S. S. J. (2015). How a Historically Black College University (HBCU) established a sustainable online learning program in partnership with Quality Matters™. *American Journal of Distance Education*, 29(3), 198–209. https://doi.org/10.1080/08923647.2015.1057440

Hsu, H.-C. K., Wang, C. V. & Levesque-Bristol, C. (2019). Reexamining the impact of self-determination theory on learning outcomes in the online learning environment. *Education and Information Technologies, 24*(3), 2159–2174. https://doi.org/10.1007/s10639-019-09863-w

Inman, J. (2021). Grounded in research: Be good, or at least evidence-based. In J. Quinn (Ed.), *The Learner-Centered Instructional Designer: Purposes, Processes, and Practicalities of Creating Online Courses in Higher Education* (pp. 69–78). Stylus Publishing.

Jaggars, S. S. & Xu, D. (2016). How do online course design features influence student performance? *Computers & Education, 95*, 270–284. https://doi.org/10.1016/j.compedu.2016.01.014

Kim, Y. K. & Lundberg, C. A. (2016). A structural model of the relationship between student–faculty interaction and cognitive skills development among college students. *Research in Higher Education, 57*(3), 288–309. https://doi.org/10.1007/s11162-015-9387-6

Legon, R. (2015). Measuring the impact of the Quality Matters™ rubric: A discussion of possibilities. *American Journal of Distance Education, 29*(3), 166–173. https://doi.org/10.1080/08923647.2015.1058114

McGahan, S., Jackson, C. & Premer, K. (2015). Online course quality assurance: Development of a quality checklist. *InSight: A Journal of Scholarly Teaching, 10*, 126–140. https://doi.org/10.46504/10201510mc

Morris, S. M. (2018). Critical instructional design. In S. M. Morris & J. Stommel (Eds.), *An Urgency of Teachers*. Hybrid Pedagogy Inc.

Niemiec, C. P. & Ryan, R. M. (2009). Autonomy, competence, and relatedness in the classroom. *Theory and Research in Education, 7*(2), 133–144. https://doi.org/10.1177/1477878509104318

Okun, T. (2021). *White supremacy culture*. https://www.whitesupremacyculture.info

Peralta District, P. C. C. (2020). *Peralta online equity rubric*. https://www.peralta.edu/distance-education/online-equity-rubric

Quality Matters. (n.d.). *About QM*. https://www.qualitymatters.org/why-quality-matters/about-qm

Ralston-Berg, P. (2014). Surveying student perspectives of quality: Value of QM rubric items. *Internet Learning*. https://doi.org/10.18278/il.3.1.9

Roehrs, C., Wang, L. & Kendrick, D. (2013). Preparing faculty to use the Quality Matters model for course improvement. *MERLOT Journal of Online Learning and Teaching, 9*(3).

Sam, J., Schmeisser, C. & Hare, J. (2021). Grease trail storytelling project: Creating indigenous digital pathways. *KULA: Knowledge Creation, Dissemination, and Preservation Studies*, 5(1). https://doi.org/10.18357/kula.149

Schunk, D. H., Meece, J. L. & Pintrich, P. R. (2014). *Motivation in education: Theory, research and applications*. Pearson.

Schwegler, A. F. & Altman, B. W. (2015). Analysis of peer review comments: QM recommendations and feedback intervention theory. *American Journal of Distance Education*, 29(3), 186–197. https://doi.org/10.1080/08923647.2015.1058599

Swan, K., Matthews, D., Bogle, L., Boles, E. & Day, S. (2012). Linking online course design and implementation to learning outcomes: A design experiment. *The Internet and Higher Education*, 15(2), 81–88. https://doi.org/10.1016/j.iheduc.2011.07.002

Thomas, L. (2015). Developing inclusive learning to improve the engagement, belonging, retention, and success of students from diverse groups. In M. Shah, A. Bennett & E. Southgate (Eds.), *Widening Higher Education Participation* (pp. 135–159). https://doi.org/10.1016/b978-0-08-100213-1.00009-3

Tinto, V. (2005). Reflections on retention and persistence: Institutional actions on behalf of student persistence. *Studies in Learning, Evaluation, Innovation, and Development*, 2, 89–97.

Wang, C., Hsu, H.-C. K., Bonem, E. M., Moss, J. D., Yu, S., Nelson, D. B. & Levesque-Bristol, C. (2019). Need satisfaction and need dissatisfaction: A comparative study of online and face-to-face learning contexts. *Computers in Human Behavior*, 95, 114–125. https://doi.org/10.1016/j.chb.2019.01.034

Winkelmes, Bernacki, M., Butler, J., Zochowski, Golanics & Weavil, &. (2016). A teaching intervention that increases underserved college students' success. *Peer Review*, 18, 31–36.

Winkelmes, M.-A., Boye, A. & Tapp, S. (2019). *Transparent design in higher education teaching and leadership: A guide to implementing the transparency framework institution-wide to improve learning and retention*. Stylus Publishing.

Siberian Syndrome in Online Learning
Impulses for utilising the experiences of Ira Shor with young adult learners

Natalie Shaw

In a milestone work of critical pedagogy, Ira Shor offered his experiences of teaching the topic of "utopia" to inner city working-class youth in his book *When Students Have Power* (Shor, 1996). Introducing the term "Siberian Syndrome" (Shor, 1996, p. 14) for the way that students acted in response to established power systems in education they had experienced so far, Shor's narrative traces the possibilities of "escaping Siberia" (Shor, 1996, p. 61) through a disruption of the power dynamics that usually frame the teacher-student interaction, casting both in mutually reinforcing roles that they find hard to escape. Shor's core motivation in attempting to break free from entrenched behaviours and teacher-student interactions lies in his concern about a 'high tide of conservativism' (Shor, 1996, p. xi), which he sees connected to the 'unreasonable order of things' (Shor, 1996, p. x): students' acceptance of their prescribed role of powerless recipients whilst purportedly being led to intellectual freedom through the education they are receiving.

In this text, I will attempt to transpose Ira Shor's experience to the enforced sudden shift to online teaching during the 2020/2021 COVID-19 pandemic, drawing on my experiences within a small teacher international training programme at a university in the Northern Netherlands. The young adult learners of this programme are undergraduate student-teachers that aspire to work in international education, and our students have a broad range of nationalities and background experiences.

After a short sketch of the background of my own online teaching experience, I will summarise some of the main points of Shor's experiences, highlighting key dynamics and insights that Shor found to be at work in his particular situation. Based on these insights, I will offer thoughts about how an escape from Siberia may be affected in educational programmes that suddenly find themselves in the situation of having to shift from face-to-face instruction to offering educational activities online. In doing so, I am fully aware of the many difficulties encountered by students who suddenly find themselves being taught via a screen, such as the lack of empathy and personal connection

149

(García-Pérez, Santos-Delgado & Buzón-García, 2016), the need for a careful look at the intercultural aspects of digital learning (Resta & Laferrière, 2015), and concerns about other dimensions of equity, both in a very material and in a sociological sense (Willems, Farley & Campbell, 2019).

My intention in writing this text is not to negate that these issues exist; nor to suggest that education is ever free of bias or fraught power relations that continue to marginalise large groups of its participants. Rather, I would like to engage in a thought experiment that takes inspiration from Ira Shor in considering whether it is possible to view an unexpected shift to online teaching as a critical pedagogical opportunity. Such a view attempts to take into account the perspective of staff (Willems, 2019) as well as students and tries to offer a path of agency for both lecturers and students in a situation that can feel disempowering.

Digi-Buddies, Infant Hybrid Pedagogies, Rotations, and Other Attempts at "Doing Online"

When the measures regarding the COVID-19 pandemic were taking effect in Spring 2020, our university programme was largely inexperienced with any form of online education. Training teachers for international schools, we believe, is best achieved through engaging in dynamic, face-to-face learning experiences that foster a sense of community, much like the sense of community we would like to see students creating in their future places of work around the globe. However, in a spirit of flexibility and with a willingness to model rising to challenges, our staff took to the requirements of mainly synchronous online learning generally more or less enthusiastically. What Willems (2019) sketches out was true for our faculty as well: large differences existed. Some staff members had extensive training and/or experiences in using technology in both teaching and learning; others embraced technology in their everyday life and felt therefore confident to attempt the use of new systems; yet other staff members were more reticent in their approach to technology, both privately and professionally. As an established and committed community of practice, we aimed at supporting one another, which occasionally led to a flurry of recommendations sent to our inboxes by savvy colleagues who were trying out the latest in terms of digital pedagogy. Other colleagues shyly mentioned that they had just discovered how to create a word cloud with the students. Our students, on the whole, were forgiving of our enthusiastic overkill in sessions where no less than 10 online tools were deployed, and equally supportive in sessions where a lecturer simply talked for 2 hours, as in a regular lecture hall, without any noticeable adaptations to the online medium.

As a faculty that embraces design thinking (see e.g., Gallagher & Thordarson, 2020) as a leading idea for its pedagogy, the overall approach we applied to the situation was that of rapid prototyping (Dow & Klemmer, 2011). We began by assigning students to one another as digi-buddies, identifying students who were willing to connect online learning peers to the classroom in real time. Efforts were undertaken to train the students willing to act as digi-buddies, as we recognised that this was a rare opportunity in honing pedagogical skills for those willing to adapt this role. This involved helping the students who were still being physically present and who volunteered to connect others digitally in how they could do so without compromising their own learning process due to being preoccupied by supporting others.

At the same time, other subject leaders chose to divide the teaching tasks, assigning some colleagues to teach the few remaining students that were physically present, whilst others guided the ever-growing group of students learning online. Both student groups arose spontaneously, with some students simply stuck in remote corners of the world and others marooned in our small Dutch university town. When it came to presence teaching, further challenges were presented by the varied room capacities on our small campus: only two rooms allowed the accommodation of a full tutor group, oftentimes the organising unit for instruction. Careful rotations were devised that would allow tutor groups to occasionally be together in a session with one member of faculty, whilst at other times being divided across different smaller rooms for pre- or post-tasks. From week to week, we tried to gather feedback from the students about their experiences and attempted to make changes where possible. When students let us know that the pre- and post-tasks were less successful because they missed the clear leadership an instructor would offer, we worked with students in higher years of the programme to provide the scaffolding the students professed to miss.

Colleagues with professional experience or training in digital pedagogy offered support. Others scoured educational publications from around the globe and updated the team on developments that were successful in other parts of the world, providing springboards for brainstorming sessions and practical solutions. Overall, as a team we performed rather well, offering both practical and pedagogical support to one another whenever this was necessary.

Toward a Critical Instructional Design

Connecting to Ira Shor: Siberians, Now and Then

In his account of the experiences surrounding teaching a university course to White working-class students from lower socio-economic backgrounds, Ira Shor sketches out a journey from framing the problematic initial situation to a disturbance of the status quo, arriving at new possibilities that were able to transcend the boundaries of the previously possible.

Analysing the initial situation, Shor introduces the term 'Siberian Syndrome' (Shor, 1996, p. 14) to illustrate students' tendencies to choose seating that is located as far as possible from the teacher. In doing so, Shor explains, students respond to what they had learned to accept as the reality of the educational system: 'unilateral authority for the teacher and a curriculum evading critical thought' (Shor, 1996, p. 13). When describing the syndrome, Shor is careful to point out that this is not due to inherent characteristics of the students but due to painful and repeated encounters with an educational system that is ignorant of, and remains uninterested in, students' personal backgrounds, aspirations, achievements, and struggles. The teacher, equipped with the institutional authority to exercise structural and procedural control concerning the syllabus, methods, and evaluation of students' work, meets students who are '[s]chooled to follow orders and to fit into lesser places' (Shor, 1996, p. 14).

Shor connects the result of students' educational 'socialization' (Shor, 1996, p. 15) with a sensitive analysis of students' emotional states in response to the experiences they have had prior to arriving in his classroom. Both, teacher and students, find themselves entrenched in roles that are constructed for them: whilst rejecting the dynamics present, students also reproduce the teacher's dominion by expecting him to behave with authority. This in turn permits students to resist in ways that have honed over long and oftentimes painful years of schooling. Shor also observes that students construct a dimension of unreality when it comes to their schooling experience, suggestive of a truer identity that is enacted in contexts beyond and far from the classroom (Shor, 1996).

Following the call to create a relationship that overcomes what is initially seen as 'antagonistic' (Shor & Freire, 1987), Shor proceeds to carefully move towards addressing the power imbalances framing the educational situation in his classroom. In doing so, Shor recognises the impossibility to do so without carefully scaffolding the process and exercising the authority of his teacher role to shape the negotiations for a more democratic approach to the course. Both student skills as well as student attitudes are obstacles to an immediate role-reversal, as Shor observes: students may resist the unfamiliar role of the teacher and the more active role this arrangement provides for them, whilst at the same time being not yet skilled enough in taking power in a way that would foster their learning process (Shor, 1996). In the words of Shor, students 'know how to follow or to frustrate authority [...], but not how to assume authority' (Shor, 1996, p. 20).

A good 30 years after Ira Shor's experiences in inner city classrooms in Staten Island, his observations still ring true of the educational reality I experienced in my present university classrooms. For each of the courses I have taught since entering the higher education teaching profession 5 years ago, I could readily name the inhabitants of Siberia: the students who prefer to hide away at the back of the room, having learnt to be sceptical of an educational system that appears to recognise ways of being and knowing that do not reflect what these students may be able to offer. Their experiences and attitudes indicate that they clearly detect the workings of a hidden curriculum not constructed in their favour (Smith, 2013) and are willing to assume the roles they are relegated to.

The oscillation between deference and resistance that Shor describes is familiar, as is the description of students who seek out the front of house seats reserved for 'the few students who share their teacher's enthusiasm' (Shor, 1996, p. 13). In my case and in that of my colleagues, the accuracy with which these observations still holds true today must concern us twofold. We should not only be concerned about the power dynamics and neoliberal education systems that faithfully reproduce these modern-day inhabitants of academic Siberia. As teacher educators, the realisation that we are working with potential 'agents of social change' (Bourn, 2016, p. 63) in the making is both exhilarating and frightening. If we succeed, we are able to foster a new generation of democratic educators that will in turn be able to affect their students. If we fail, we may have to face the realisation that we are yet another cog in the wheel of an educational machinery that generates new blueprints of the status quo for generation after generation.

Siberia 3.0: Tracing Patterns of Siberian Syndrome in Online Learning

Reflecting upon the experience of nearly a full academic year of more or less online teaching activities, it is startling to realise how readily

Siberia appears to have migrated online in my teaching context. In an environment that is both conceptually and spatially removed from the regular classroom (Jones & Lloyd, 2013), patterns of familiar behaviour persist with surprising regularity. Online, the front row students have transformed into those vocal in chat functions when encouraged to comment; those willing to raise their virtual hand and volunteer a comment; those happy to put on their camera to signal their presence. The back of the physical room, it appears, translates to silent attendance in digital spaces, a blank screen, and sparse participation in discussions, be they conducted through voice contributions or via the chat. Siberian Syndrome students may request for lessons to be recorded and opt for the safety of having the content represented without the danger of being asked to participate more actively—after all, the back of the physical room allows a certain safety not afforded by a list of participants in the online environment, where anybody may be called upon at any time.

The distance may be explained by differences that persist in the real world and that effectively contribute to the creation of Siberia in the first place. Young people from academic homes of the middle classes most readily display the habitus of 'eager student' in the analogue classroom and feel comfortable in the presence of the teacher, whilst those whose backgrounds may have prepared them less for the interactions in academia shy away from too personal a contact, afraid of having their perceived lack of understanding revealed, suffering from social anxieties, or having other reasons for being more withdrawn. During the pandemic, many students resided with their families. Thus, in digital spaces, the literal background of someone's on camera presence may have been felt by students as indicative of their socio-economic and cultural background as a whole. Even prior to the pandemic, Gilliard (2017) offered an eloquent and lucid criticism of the workings of capitalism and surveillance in digital spaces that lead to a marginalisation within our student population.

Put placatively: The front-row inhabitants most often revealed themselves sitting in front of richly furnished bookcases, reportedly sitting in parents' offices or other work furnished with the signifiers of a middle class life sharing affinity with academia. The online Siberians most often preferred to not present on camera. On occasions where an on-camera presence could not be avoided, other spaces became visible: more cramped housing, others sharing the same space, fewer signs indicating 'academia'. To be clear: neither cramped housing nor the presence of others in a learning space necessarily indicate that a household may not be academically versed. However, my point is that many students might have concluded, based on a perception of still active hidden curriculum workings, that their backgrounds (both literal and figurative) did not measure up to a somewhat nebulous, but nevertheless persistent standard. And, to be fair: amongst colleagues co-teaching the same courses and witnessing students' on-screen backgrounds, we did occasionally comment on the literal backgrounds of some students. Who is to say how many evaluations have entered our subconscious, influenced by a pernicious bias that our transformatively engaged, conscious teacher selves would prefer not to have? Who is to say how many digital Siberians perceived this with clarity and preferred to not expose themselves further to our evaluating gaze?

As Shor (1996) suggests, these responses are not without solid grounding: rather, they speak about students' prior experiences with a reality that, whilst professing to be inclusive, often does not escape its inherent biases. As Zambito (2020) argues, bias in online learning is most effectively confronted by instructors willing to acknowledge their own biases. In my experience, this opportunity arose only sporadically, with individual colleagues reflecting upon their perceptions of students' engaged presence, perceived apathy, or their personal responses to the students' visible backgrounds during casual professional exchanges. As a faculty, more learning opportunities could have been taken in reflecting more structurally about the effects these perceptions had on the framing of the teaching and learning situation in the initially unfamiliar online realm.

Arrival times may also indicate a degree of Siberianism: whether presence learning classrooms or in the digital space, front row students tend to arrive early, eager to share a few minutes of conversation with the teacher. Late arrivals oftentimes populate the back spaces of the classroom, apparently content to slip in just before a session gets underway. In our digital spaces, the same students arrived silently or late, and thus missed out on the interactions that maintain a semblance of connectivity in times when real-life encounters are few and far between.

When Life Hands You Lemons: "Untested Feasibilities" in the Digital Realm

Shor reflects upon the difficulty of disrupting routines that are ingrained after a lifetime spent as students in the educational system and seeks to locate 'untested feasibility[ies]' (Shor, When students have power: Negotiating authority in a critical pedagogy, 1996, p. 3). This concept, adopted from Freire (Shor & Freire, 1987), describes the experiences that may be possible when educators extend the boundaries of the present and attempt to widen the scope of possibilities together with their students.

Many educators worldwide have struggled with making the unexpected pedagogical and technological shift towards online teaching (Willems, 2019). The sudden migration to online learning has once more revealed equity issues surrounding not only access to the necessary infrastructure and digital literacy (Ragnedda, 2020), but also the equity of digital content itself with regards to culture, bias, and language (Willems, Farley & Campbell, 2019). However, what I would like to suggest is that we may also be framing the experience in a new light. What I would like to put forward is the possibility to view the temporary migration to digital learning as a rare opportunity, an 'untested feasibility' in the Freirean sense.

As such, this suggestion is reminiscent of the adage that the lemons of having to rapidly adapt to a global pandemic may be viewed as lemonade in the making. Transposing the saying to the educational context, the following design question arises: How might we harness the possibilities offered by a migration to online learning whilst retaining a critical awareness of the barriers that persist with regards to equal access and content in the digital realm?

When Jones and Lloyd (2013) remark on the absence of a sense of physical space in online learning environments, this remark seems to offer an implicit invitation to seize upon the absence of physical space as a disruption of the usual patterns of teacher-student interaction. Naturally, the power dynamics that students are aware of and that Shor (1996) and Shor and Freire (1987) acknowledge as an inescapable fact of the educational context still persist. Students are well aware of the fact that teachers retain the power to assess their learning and to award grades at the end of the course. Yet other well-honed tropes may at least be minimally disrupted, and afford us an angle of attack for widening the cracks.

When Shor (1996) writes 'As the teacher, I'm supposed to go to the front and assert the authority vested in my position' (ibid., p. 16), it does not take much imagination to transpose this action to what students expect in an online environment. Naturally, the expectation is that the teacher or lecturer will take the lead in the educational process, outline the aim of the lesson, structure the process, and demarcate the fault lines between acceptable and unacceptable behaviour. However, utilising the question posed by Jones and Lloyd: 'Which way is up?' (ibid., in the title), we may present students with a similar question: Where, exactly, is the front from which they expect us to lead? How does this physical space translate to actions in the digital world? What expectations do they usually have of a teacher, and how may these differ when learning happens online? The shift from physical to online spaces, if utilised, may serve to facilitate the delicate process that Shor (1996) outlines in his book: the careful negotiation of a more democratic curriculum in which power, agency, and responsibility are shared and established between the participants of the educational experience.

As lecturers and students, we brought and still bring different skills and understanding to our shared learning situation. Whereas we educators possessed procedural expertise, planning and mapping out the learning trajectories, students supplied expertise of a different kind: pragmatic understanding of the situations of their fellow students and oftentimes very effective solutions to complex problems of connectivity that were rooted in their more complex understanding of how students are connected amongst one another. On occasion, students have proven more skilled than us lecturers in navigating some of the technology available to us, reminding us that the role of expert carries many facets: knowledge expert, procedural expert, skills expert, and so forth. This realisation alone can function as an invitation to critically reflect with students upon these facets of expertise and foster the realisation that lecturers may not always be experts, and students may well have expertise in areas that they had previously not recognised as areas of strength. This realisation itself can work as a springboard towards establishing the 'dialogical relationship' (Shor & Freire, 1987, p. 92) that is at the heart of Freire's democratic vision.

Further, as Shor (1996) poignantly illustrates, the disruption of the teacher–student relationship in an actual classroom is not without its discomforts. The physical discomfort when Shor situates himself amongst the students of Siberia culminates in students removing themselves from the vicinity of the teacher and choosing seats further away from Shor's embodied authority. In an online learning situation, it is possible to disrupt in a way that appears somewhat gentler, less physically confronting, and yet possibly still as effective. The equivalent to Shor's repositioning may take the form of a carefully scaffolded stepping back of the lecturer from shaping the lesson to her agenda. Alternatively, it may be orchestrated as a lecturer choosing to join the breakout groups where inhabitants of 'deep Siberia' (Shor, 1996, p. 28) are gathered, and listening without taking charge of the discussion.

When students begin to reflect on the legacy of time spent in the educational system that provides restrictive roles for the marginalised, this process can elicit strong emotions. Anger, frustration, bitterness, and sadness may result from a consideration of how their own position at present has been influenced by societal workings beyond their control; all the while holding them accountable via the neoliberal trope of their own responsibility. Zembylas (2012) warns that critical pedagogy may be in danger of overlooking the significance of emotion and suggests ways of working with emotional knowledge without avoiding 'pedagogic discomfort' (Zembylas, 2012, p. 8). When educators scaffold reflection on power imbalances and equity as a precursor to dialogical relationships with students, it may be argued that doing so in physically present classrooms allows for a more empathic approach of recognising the effects this subject matter has on students. On the other hand, young adult learners can easily resent being put in a vulnerable position in front of their peers and may welcome the distance of a well-scaffolded online approach, where discussions and reflections are elicited and then carefully followed up on without the immediacy of a physical classroom environment where strong emotions may make students uncomfortable in front of others.

In Conclusion

In moving education from a physical to a digital plane, we have not left behind any of the issues that have plagued educational systems for generations: a neoliberal drive for efficiency, a narrowing of curricula, a heightened focus on outcomes at the expense of providing space for critical thought and dissent, a lack of cultural responsivity. However, we may have gained a unique opportunity in forcing open the miniscule cracks that emerge when education suddenly and unexpectedly migrates from presence to online learning.

Opportunities may present themselves in reconsidering and problematising the ways of being in the physical classroom from the vantage point of the online platform, an unfamiliar situation less fraught with finely-honed patterns of student-teacher interaction. The online learning environment is far from a power-free vacuum where everything is possible. However, we may feel invited to seize the opportunity to question some of the behaviours we usually associate with the role of teachers and students and examine them afresh. Where, figuratively, is the front of the classroom in the online space? What behaviours do we associate with leading from the front? How might we transform these into the beginnings of a democratically negotiated curriculum?

Not all these disruptions have consistently happened or been seized upon in my own experience with online teaching. Initially, the temptation to simply put something out there that would keep students gainfully occupied towards achieving the learning outcomes was oftentimes the best we could produce in a short amount of time. However, the reflections presented here are certainly an inspiration towards framing moments of disruption not as an inconvenience, but as a unique and rare opportunity to upend the status quo in a more natural way than the path that Shor (1996) sketches out.

One big realisation is that the sudden shift to online learning has exposed ourselves to one another in quite unexpected ways. As a learning community, we have seen one another's living rooms, kitchens, sofas, desks, wall decorations, family photos, pinboards, partners, dogs, cats, and once, memorably, rabbits. We have jointly laughed at hilarious incidents when microphones were left unmuted to reveal a student singing loudly, and marvelled at moments such as a student remarking 'Natalie, I think there is a policeman outside of your houseboat' (To clarify: the officer in question was trying to attract my attention to question those in the neighbourhood about a break-in in a nearby shop that had occurred the night before). Jointly, we have shared anecdotes such as the memorable attempt of a grandparent repeatedly trying to summon their grandchild to lunch by calling their name loudly—all whilst the student in question was presenting their final work to us via Teams at the same time.

As lecturers and students slowly return to the face-to-face learning situation we are accustomed to, a big question remains. How will we transform the fact that we all gained immediate insights into one another's lives into a cornerstone of more equitable, culturally responsive teaching opportunities, as suggested by Hammond (2015), into a more democratically negotiated classroom as advocated for by Shor (1996)? Of course, we could collectively choose to pretend that the past years of the pandemic and its sudden shift to online teaching had been nothing more than an aberration, and ignore the insights that we had gained into one another's lives, the moments where we suddenly found ourselves relying much more forcefully on one another's expertise and knowledge.

To be sure: not all members of our faculty are equally comfortable with regarding the miniscule cracks that have opened up in our usual roles of lecturer and student as worth expanding and widening. Some may be grateful to return to an established routine where everyone plays their clearly mapped-out role. However, most of us are at least comfortable with opening up dialogue around this shift: What has this time taught us, collectively? How can we harness these changes to arrive at a classroom that is more equitable? Most importantly, what do students think about this, and how would they like to see us continue together? As Shor (1996) suggests, we may specifically choose to sit amongst the students once again huddled at the back of our physical classroom to ask these pivotal questions.

Rather than having to create a situation akin to the Freirian concept of an 'untested feasibility' (Shor, 1996, p. 3), it literally fell into educators' laps. For those amongst us, myself included, who feel that they have not yet maximised the opportunities, it is not too late. Stepping back into physical classrooms after the end of the COVID-19 enforced period of rapid-development online learning presents another opportunity for disrupting patterns of Siberian Syndrome and disturbing the status quo.

References

Bourn, D. (2016). Teachers as agents of social change. *International Journal of Development Education and Global Learning*, 7(3), 63-77.

Dow, S. P, & Klemmer, S. R. (2011). The efficacy of prototyping under time constraints. In H. Plattner, C. Meinel, & L. Leifer (Eds), *Design thinking: understand – improve – apply* (pp. 111-130). Springer.

Gallagher, A. & Thordarson, K. (2020). *Design thinking in play: an action guide for educators*. ASCD.

García-Pérez, R., Santos-Delgado, J.-M. & Buzón-García, O. (2016). Virtual empathy as digital competence in education 3.0. *International Journal of Educational Technology in Higher Education*, 13, 13-30.

Gilliard, C. (2017, July 3). Pedagogy and the logic of platforms. *EDUCAUSE Review*. https://er.educause.edu/articles/2017/7/pedagogy-and-the-logic-of-platforms

Hammond, Z. (2015). *Culturally responsive teaching and the brain*. Corwin.

Jones, D. & Lloyd, P. (2013). Which way is up? Space and place in virtual learning environments for design. *DRS Cumulus Oslo 2013 – Proceedings from the 2nd International Conference for Design Education Researchers*, 1, 552–563.

Ragnedda, M. (2020). *Enhancing digital equity: connecting the digital underclass*. Palgrave Macmillan.

Resta, P. & Laferrière, T. (2015). Digital equity and intercultural education. *Education and Information Technologies*, 20(4), 743-756.

Shor, I. (1996). *When students have power: Negotiating authority in a critical pedagogy.* Chicago University Press.

Shor, I. & Freire, P. (1987). *A pedagogy for liberation: dialogues on transforming education.* Bergin & Garvey.

Smith, B. (2013). *Mentoring at-risk students through the hidden curriculum of higher education.* Lexington Books.

Willems, J. (2019). Digital equity: considering the needs of staff as a social justice issue. *Australasian Journal of Educational Technology*, 35(6), 150-160.

Willems, J., Farley, H. & Campbell, C. (2019). The increasing significance of digital equity in higher education: An introduction to the digital equity special issue. *Australasian Journal of Educational Technology*, 35(6), 1-8.

Zambito, V. (2020, September 11). Does bias exist in online learning? Yes, but it doesn't have to. *eLearning Industry.* https://elearningindustry.com/what-need-to-know-about-bias-in-online-learning

Zembylas, M. (2012). Critical pedagogy and emotion: Working through 'troubled knowledge' in posttraumatic contexts. *Critical Studies in Education*, 54(2), 1-14.

Building from Our Sensitive Edges
Transgressive strategies for dialogic online course design

Meryl Krieger and Clayton D. Colmon

> I wonder what our sensitive edges have to teach us. What do our mortality and openness mean to the ecology we could surrender to together?
>
> – Alexis Pauline Gumbs

This collaborative work centers marginalized, "non-traditional," adult learners within a thick discussion of teaching and learning. While preparing to launch an online certificate program, we noticed the discursive energy that's often spent identifying the needs of "traditional" student populations.[1] In response, we've decided to redirect this energy through practices that open us up to a broader range of needs for growing digital learning communities and that embrace our sensitive edges. What do we mean by "sensitive edges"? We are most aware of the need for adaptability and openness to how students are engaging with materials, how they are creating community in a class space, and ways that good teaching can guide them towards ownership of their own learning. We use Alexis Pauline Gumbs' shorthand of "sensitive edges" to reference these three goals (Gumbs, 2020). Teaching with your sensitive edges allows students to make their own connections to materials. bell hooks and Paulo Freire help us to articulate a theoretical praxis that we can use in a systematic and reflective way to grapple with large and thorny concepts.

Our efforts speak to some of the possibilities and challenges within instructional design praxis while offering a path for change in online

---

[1]. Many define traditional learners as students aged 18–24, who are attending college or university full time, and who have entered college or university directly from completing their high school (secondary) education. In contrast with traditional learners, non-traditional "adult learners" are students who are returning to school settings some time after the completion of their high school (secondary) education. See in particular data reported by the U.S. Department of Education, National Center for Education Statistics, including characteristics by student population, at the end of this chapter, for further reading. This data set is regularly updated

course design through a manifesto that offers strategies for transgressive work. A manifesto in its most essential form is defined as "a public declaration of objectives, opinions, statements or motives" (Dictionary.com). Building on that definition, we see this chapter as a collaboratively written collection of questions and responses that creates space for larger conversations. In this chapter, we begin by framing the theories and concepts that formed the framework for the collaboration which created the Digital Strategies and Culture certificate (DIGC) within the Penn Liberal and Professional Studies Online (PLPSO) program, followed by a deeper case study analysis of the certificate program itself.

Co-Conception: Manifesto and Metalogue

In framing his work *Radical Hope: A Teaching Manifesto*, Kevin Gannon notes "It's easy to critique but harder to build. Yet we owe it to ourselves and our students not only to point out the vast array of problematic areas in the higher educational landscape but also to offer tangible and meaningful alternatives." (Gannon, 2020, p.12) As a complement to this, Gregory Bateson gives us the framework of a metalogue which he describes as a type of conversation about an idea or subject. For Bateson, "[t]his conversation should be such that not only do the participants discuss the problem but the structure of the conversation as a whole is also relevant to the same subject." (Bateson, 1987, p.12) We combine these two approaches in this chapter. We use the concept of a manifesto to create a call to action that recognizes the needs and agency of adult learners through critical instructional design methods.

Both the writing process for this chapter and our approach to designing DIGC have followed Bateson's idea of the metalogue. This complements the goals we have for the online certificate program and the discussion of that program we present here. We see it necessary to distinguish online learning as a space for transformational education that's accessible, open, and ultimately adaptable to student needs; we affirm the modularity and flexibility that's built into our courses as one example of an approach that honors the value and concerns of adult learners; we ground the larger degree program in the complex realities that non-traditional/adult learners face in higher education—which helps us humanize learners and their experiences;[2] and we, ultimately, af-

---

2. Friere tells us the importance of our humanizing "preoccupation" with the beautiful messiness of emergent critical instructional design. A number of key ideas in this chapter are grounded in the Friere method—empowering people to take control of their lives through literacy education, which he considered necessarily anthropological and contextual to the cultural settings

firm mutualistic spaces for shared learning online that disrupts older, traditional models that don't privilege learners' needs.

What does it mean to support mutual accountability between instructors and learners?

Mutual accountability is an outgrowth of meaningful learning. It involves creating spaces for learners and instructors to share the role of authority and expert. It can also mean questioning what authority and expertise look like in core course readings and materials as well as in knowledge-building work outside formal course spaces. Traditional pedagogical learning environments cede this role primarily to the instructor. In contrast, andragogical methods recognize the importance of sharing that authority with adult learners whose experiences, concerns, and wisdom can support contextualized knowledge-building. This approach honors the practical realities that our students shape in each course.[3]

Building on Gannon, the DIGC certificate program establishes transgressively hopeful critical instructional design practices in "elite" academic spaces—to which many non-traditional students have had unequal access. He notes, "if higher education is indeed the social and political good we believe it is, then we should be doing our level best to ensure as many students as possible are able to access the opportunity to pursue it" (Gannon, 2020, pp. 73). This involves all areas of higher education—even within institutions whose history of prestige and power are built on exclusivity, fixedness, and inequality. Gannon shows us how to frame access and flexibility as components of radical hope. Both he and hooks affirm the many ways non-traditional learners contextualize their presence and participation in learning environments. We argue that this extends to the critical instructional design work that informs these environments. In the DIGC certificate, we build on the explicit goals of inclusion and equity in the University of Pennsylvania Liberal & Professional Studies Online Bachelor's in

---

in which learners live. Friere developed a dialogic pedagogy in which learning and critical thinking were co-created by leaders and learners—where "the method ceases to be an instrument by which the teachers... can manipulate the students ... because it expresses the consciousness of the students themselves... Consciousness is thus by definition a method, in the most general sense of the word" (Freire & Macedo, 2021, p. 66).

3. Following the guidance of bell hooks and Audre Lorde, we wrestle with what it means to support interdependence and accountability via the work of teaching, which is also the work of survival "no matter where we key into it."

Applied Arts and Sciences degree program[4] and, with the DIGC certificate's focus on critical thinking and literacy in the context of contemporary digital culture, privilege the voices of our learners.

How do we frame forms of inclusion and equity for non-traditional student populations who embody and demand different pedagogical narratives?

Historically there has been less diversity and inclusion for non-traditional student populations in elite, undergraduate programs. For some, a shift away from this model requires rethinking what exactly "elite" means and how it impacts higher education as a public good. Community colleges, city colleges, regional colleges and commuter-focused institutions have specialized in addressing the needs of nontraditional student populations for generations. Breaking from the traditional perspective of an Ivy League American university, PLPSO ultimately takes critical steps toward creating an adult learner-focused community that reflects the changing demographics of student populations, and the shifting needs of the learners themselves.

Friere notes that through the use of dialogue, or, in his terms, dialogic pedagogy, teachers and students "become jointly responsible" (Freire & Macedo, 2021, p.74) through what Freire calls "problem-posing education," where "people develop their power to perceive critically *the way they exist* in the world *with which* and *in which* they find themselves; they come to see the world not as a static reality, but as a reality in process, in transformation" (Freire & Macedo, 2021, p.77). Freire emphasizes what he called "praxis" which holds that action and reflection on the world is necessary to change it (hooks, 1994, p.14).

Access and inclusion in our virtual classrooms leads us to think about the importance of traditional and human-centered learning design. Traditional learning design approaches often center preordained outcomes, prefabricated rubrics, and prescriptive visions of human possibility in educational spaces. We see examples in methodological mainstays like Bloom's Taxonomy, the ADDIE model, the SAMR Model, and others that encourage designers to manage learning through tidy and efficient practices that privilege replicability over contextualized care. While efficiency can save time, it's too often weaponized to deval-

---

4. "The specific needs of that adult learner audience drove the creation of Penn LPS Online, developed to further advance Penn's mission of making a relevant, high-quality education accessible and affordable for working adults. " From (Penn LPS Online, 2018)"

ue and flatten the labor of teaching and learning in service of capitalist economies of scale that rely on rigid, extractive temporalities.[5] This fact is particularly consequential for folks whose devalued labor requires them to teach too many courses—rostered with too many students—in too many places. But what would it look like to build practices that move us closer toward transformative humanization while speaking to the realities of precarious, time-sensitive, labor and de-agencied learning? As Kevin Gannon notes, "if we want to restore the idea of higher education as a space for transformation, of emancipatory learning, then we need to start with the ways in which we talk about its purpose and value" (Gannon, 2020, pp. 110).

When we began designing the DIGC certificate, we started with the question: what if we purposefully designed this transformative space to center and care for human learners in ways that exposed the dehumanizing limits of learning management systems? For our purposes, at its best, human-centered design (HCD) invites us to think critically about choices that affect how we build and maintain learning spaces for folks. It also creates room for us to reflect on the design choices we make for learners within those spaces. Recent iterations of HCD have responded to calls for contextual, inclusive, and equitable design by embracing practices that help ensure equity and social justice for historically marginalized learners. This includes intersectional considerations of race, gender, sexuality, disability, and neurodivergence. Still, some HCD work continues to perpetuate narrow visions of *which* humans should be centered, what aspects of their humanity should be affirmed, and how much room their sociopolitical presence should occupy in learning design.

bell hooks models a transgressive approach to mutualistic participation in transformative teaching and learning that combats undemocratic exclusion perpetuated in some human-centered design work. She writes, "[m]aking the classroom a democratic setting where ev-

---

5. See Rasheeda Phillips' provocations about linear and non-linear constructions of time in "Organize Your Own Temporality: Notes on Self-Determined Temporalities and Radical Futurities in Liberation Movements." She notes that "[r]adical liberation movements reappropriate notions of time and temporality itself, stealing back time to actively create a vision of the future for marginalized people who are typically denied access to creative control over the temporal mode of the future, and redefining that future's relationship to the past and present." (Phillips, 2016, pp. 49) This encourages us to ask: is critical instructional design about the business of radical liberation in its "practice of freedom?" If so, how do we use transgressive construction of time to advance this work—in the face of linear narratives about progress which often silence histories of systematic disenfranchisement and capitalist extraction?

eryone feels a responsibility to contribute is a central goal of transformative pedagogy" (hooks, 1994, p. 39). For her, transformation includes forms of solidarity, liberatory care, and freedom that demand capacious definitions of humanity. All help constitutes transgressive pedagogy that, in Freire's words, embraces all humans as "beings of praxis" who are "capable of changing the world" even as we work to "give it meaning" (hooks, 1994, p. 48).[6] Indeed, she notes that "[t]o have work that promotes one's liberation is such a powerful gift that it does not matter so much if the gift is flawed" (hooks, 1994, p. 50). Liberation can take many possible forms in education. We see critical instructional design as one of those forms and have centered it in the DIGC certificate's focus on critical digital literacy.

We frame critical digital literacy not only as a theoretical concept, but as an actionable lens from which all course participants co-create knowledge. Working with, and through, Alexis Pauline Gumbs' concept of "sensitive edges" also helps us understand critical digital literacy as a site for careful reflexivity, interdependent presence, and democratic participation—each of which shape the ecology of all DIGC courses. For Gumbs, and for us, literacy includes a version of vulnerability that makes room for the needs, perspectives, and desires of each being in a co-created space—regardless of their role. While Gumbs' explores the liquid environments of marine mammals to locate Black feminist lessons about life, love, and literacy, we turn to digital spaces in which these interwoven lessons can inform the ways we listen and build language for our critical instruction design work. Building from our sensitive edges invites us to ask: "[w]hat could it mean to be present with each other across time and space and difference" (Gumbs, 2020, p. 67) while considering the layers of literacy and communicative care such interpersonal presence requires in digital spaces.

---

6. We feel it is important to acknowledge that hooks' identity as a Black feminist is crucial to understanding both her writerly and teacherly perspective. The dialogic structure she uses in *Teaching to Transgress* gives her room to explore a reflexive praxis that upends some of the distancing strictures of traditional academic writing. It also allows hooks to offer a care-filled critique of Friere's "sexist language" and "phallocentric paradigm of liberation" in his earlier work where—she argues—"freedom and the experience of patriarchal manhood are always linked as though they are one and the same" (hooks, 1994, p. 49). She does not detach, downplay, or compartmentalize the messy bits of patriarchy in Freire's early writing. She, instead, places her reading in relation to the nuanced, personal, ways of knowing she develops with and through his work, as she shares: "there are many other standpoints from which I approach his work that enable me to experience its value, that make it possible for that work to touch me at the very core of my being" (hooks, 1994, p. 49).

Communication scholar Paul H. Arnston contributes to this conversation through his articulation of agency available to all participants in any community of discourse through distinctions of the levels to which participants are enculturated to engage. Arnston presents three roles through which people can engage in interpersonal communication with others—that of the citizen, where "individuals assume responsibility for their well-being and expect to participate non hierarchically in decision-making groups…" (Arnston, 1989, p. 32) who "evaluates the professional-client encounter based on the accomplishments of [relevant] tasks" (Arnston, 1989, p.33); the client role, where compliance with someone else's instructions or guidance is the primary relationship; or the consumer role, where the individual seeks out specific amenities in the context that determines their satisfaction with the service they are purchasing. We argue that Arnston identification of the choice of agency—what Arnston identifies as the role of the citizen—supports mutual accountability and democratic participation in the critical teaching and learning praxis (Arnston, 1989). Arnston's concept of the citizen also enables us to address issues of liberation and ameliorate the problem of exclusionary practices in traditional learning design. This idea of citizenship merged with our focus on digital literacy and competence as a pedagogical goal of the DIGC certificate program. This idea of citizenship is a node of connection—with it, faculty and students can co-create, partner, and set expectations for a class environment for how knowledge is built in a course, on a specific topic.

Faculty and students are allies, not adversaries, in the collaborative creation of knowledge. What does it mean to set expectations collaboratively, and why should we do this? Can we as educators remove limitations around student learning by simply asking them what helps them to learn—and how might this question advance education as a practice of freedom?

We often talk about education as a place to prepare for the workforce, but it's also important to consider how the skills needed here also apply in broader public spaces. The goal isn't to make learning spaces that are only self-contained, but to be a microcosm of what could and—we hope—will be in the larger world. Freire tells us that working collaboratively with learners helps them take ownership of their learning. Identifying obstacles and removing them from the learning environment makes the online course accessible and adaptable. This is especially important for adult learners who primarily come to the educational setting to address real life problems and goals rather than focusing on the acquisition of knowledge for its own sake. Arnston gives us a positionality of citizenship from which we can take ownership of our roles of educator and learner. It can also be helpful to grapple with challenges together because it gives educators and learners alike perspective on what the challenges are and how we can solve them. We can all see pieces of a solution independently, but it can take the village to put the puzzle together. All get to practice the kinds of interdependence that they will take to settings outside the learning environment.

As we approached the development of a fully online, primarily asynchronous certificate program on digital culture for a non-traditional learning population, we confronted yet other elements that seem important to unpack: traditional learning design is focused on the learning needs and strategies of traditional-aged students, using pedagogical approaches grounded in developmental psychology. Indeed, fundamental to the application of hooks and Freire to adult, non-traditional education requires problematizing basic ideas about the term pedagogy itself. Pedagogy is one of those terms educators use that comes with a vast assortment of baggage—mostly because the definition explains very little about what pedagogy, in a word, is. The Oxford Languages Dictionary, for example, begins by defining pedagogy as "the method and practice of teaching" and then qualifying this with a focus on "academic subject or theoretical concept." This definition leaves assumptions about who is being taught, and what kinds of learning structures are being used, but we can draw from lived experience that the learner population are traditional-aged students, and the model being applied is face-to-face classroom instruction.

Malcolm Knowles' two distinctions help us bridge this shift from traditional learners to adult, non-traditional learners. First, pedagogy assumes that teachers take responsibility for—and direct—learning. Second, pedagogy is focused on the education of children (Knowles, 1970). He makes the further point that pedagogy holds that knowledge is a constant. Citing Alfred North Whitehead, Knowles definition also suggests that, "...what a person learns will remain valid for the rest of his life." However, Knowles points out that major cultural changes have accelerated since the early 20th century, ensuring that knowledge gained in childhood is often no longer relevant by mid-adulthood. He concludes that lifelong learning must consequently become a focus for educators. From this concern, the term andragogy has evolved, to focus on the art and science of helping adults learn.

Finally, Knowles' understanding of andragogy posits four assumptions about adult learners and their characteristics that are distinct from ones educators might hold about child learners. These assumptions become fundamental to our ideas about how knowledge can be co-created by educators and learners and what boundaries we can reasonably set on the work expected of both partners in the learning process:

- As a person matures, their self-concept moves from dependence on others towards self-direction
- As they grow they accumulate a reservoir of experiences that become an increasing resource for their learning
- As individuals, their readiness to learn becomes oriented increasingly towards the developmental task of their social roles
- Their temporal perspective changes from learning about a subject to focusing on solving problems (Knowles, 1970, p.55)

Andragogy as a method thus becomes foundational for our work as educators and in the DIGC and LPSO program in particular. It also gives us strategies for engaging in our work as educators—that we can build on these needs and interests as mechanisms for integrating, synthesizing, and applying new knowledge.

One other element we have found continually important to center in our exploration of teaching non-traditional adults in the DIGC program is the role and impact of trauma on both learners and educators within the classroom. While this concern is relevant for all populations, it is particularly crucial to consider in the creation of learning spaces for non-traditional learners. Trauma-informed care has been defined by the Substance Abuse and Mental Health Services Administration (SAMHSA):

> TIC [trauma-informed care] views trauma through an ecological and cultural lens and recognizes that context plays a significant role in how individuals perceive and process traumatic events, whether acute or chronic. TIC involves vigilance in anticipating and avoiding institutional processes and individual practices that are likely to retraumatize individuals who already have histories of trauma (SAMHSA, 2015, p.1).

Indeed, in her 2021 article on trauma-informed andragogy, Jennifer Davidson makes the point that "a trauma-informed andragogy for the graduate theological classroom must recognize the pervasiveness of trauma and the likelihood that our students (and our colleagues, and we, ourselves) bring the experience of trauma with us into the classroom"(Davidson, 2021, p.12). Davidson's point—that we must *assume* trauma as a consideration in adult, non-traditional learning has always been a reality—is important to include in our praxis.

The idea of trauma-aware, or trauma-informed, teaching practices has gained in attention and popularity since the start of the COVID-19 pandemic. While this is important as a step in addressing the social justice concerns around traumatic events, what is more important from our perspective is to recognize that according to the National Council for Behavioral Health, 70% of adults in the United States have experienced some type of traumatic event at least once in their lives. This makes trauma-aware teaching practices even more essential, since pedagogical approaches operate on the assumption that students operate at a baseline level that is not traumatized. Indeed, it takes Wartenweiler's points that "acknowledging the impact of trauma on learning is of great importance if we want to create more socially just education systems and not disadvantage traumatized learners" (Wartenweiler, 2017, p.96). Additionally, it helps us see that it is the *majority* of students, particularly when they are adults, who have experienced trauma, meaning that the sensitivity that trauma-aware practices bring is no longer just a good idea, but a necessary component of effective education.

With all these tools in mind, we come to the essential framework of critical instructional design as the praxis where the DIGC certificate is situated. For us, critical instructional design is a practice and process of freedom that relies on intimacy, vulnerability, interdependence, democratic participation, iteration, and hope; it pushes us to lead with and through our sensitive edges;[7] It is messy at times because it centers the humans that give it shape and meaning; it is rooted in reflexivity; it neither has nor espouses best practices because all bests are contextual; it exists outside and, at times, counter to learning management systems; it is educational care work that operationalizes love; it is what

---

7. Alexis Pauline Gumbs' asks us to consider "what our sensitive edges have to teach us" in her work *Undrowned: Black Feminist Lessons from Marine Mammals* (2020) which contributes to a larger "Emergent Strategy Series." She invites us to make connections between forms of listening, being, and doing that reach across species and ecologies, arguing that "[l]istening is not only about the normative ability to hear, it is a transformative and revolutionary resource that requires quieting down and tuning in." (p. 15) We suggest that our sensitive edges allow us to listen and "tune in" across digital distances and spaces.

Adrienne Maree Brown might describe as an "emergent strategy;" and it is change (Brown, 2017). This contrasts traditional instructional design practices which often center a different, systems-thinking approach to teaching and learning.

Instructional Design's reliance on systems-thinking speaks to its early uses in military and technical training.[8] This history continues to shape some of the language Instructional Designers use to describe and imagine learning environments (by "assessing learning targets," "codifying effective objectives," etc). We also see influences of this reality in instructional design models that privilege replicable systems—no matter how "agile" the utilization.[9] While these models offer important context for traditional ways of communicating prevailing approaches to instructional design praxis, critical instructional design asks us to imagine students outside of homogenizing traditions by:

- Centering Accessibility as an iterative, action-oriented process
- Interrogating assumed hierarchies between instructors and students
- Questioning implicit and explicit biases that are encoded in tools and systems
- Situating design practices within discussions of social and transformative justice
- Decentering white, male, heteronormative voices within shared materials
- Developing a teaching and learning infrastructure that's built on care
- Fostering critical play and curiosity, even when it pushes us to and through unexpected places

These practices form the basis for the expectations and values we identify both in the original proposal for the DIGC certificate, and in the formative questions we bring to organize the DIGC courses.

---

8. See Robert A Reiser's "A History of Instructional Design and Technology: Part 1: A History of Instructional Media," and "A History of Instructional Design and Technology: Part II: A History of Instructional Design" which grew out of his earlier work on instructional technology.

9. Some of these models include, ADDIE (which we reference earlier in the chapter), SAM, Cathy Moore's Action Mapping, Sick and Cary Model of Design, Merrill's First Principles of Instruction, and a host of others.

Educators like Ruja Benjamin[10], Sasha Costanza-Chock[11], Bryan Keith Alexander[12], and others invite us to engage with marginalized learners whose presence and histories can problematize assumptions about which human experiences matter in teaching and learning. Alexander's work, for example, helps us explore how intersecting racial and sexual identities can queer human-centered design praxis. He writes "when speaking about social issues in the classroom, I must address the political potency, the psychic disturbance, and the potential physical impact of those issues on my black gay body." He goes on to reflect that: "the classroom is a space in which the personal is magnified, not diminished." (Alexander, 2005, p.251) But what happens when the personal is universalized in this magnification—via design decisions that overdetermine white, male, cis-hetero, hierarchical ideals as the default setting within teaching and learning spaces? How do we de-center this form of privileged humanity without diminishing the power of personalized/personal teaching and learning within digital learning communities?

What are the distinctions between face-to-face and online learning, and why are these important?

A traditional learning environment is, by common definition, a face-to-face classroom space in which a co-present teacher or teaching team organizes and directs student learning. This environment as-

---

10. See Ruja Benjamin's reflection on what it means to think critically about design as a concept and practice in her work *Race After Technology*. She asks readers to consider the ways design acts as "a colonizing project...to the extent that it is used to describe anything and everything" without contextualizing the many systems and "[g]eneologies [that] reflect and reproduce power relations" (Benjamin, 2019, p. 176 -177). Among the many questions she raises in the work, one sticks with us, as we grapple with critical instructional design praxis: "If design is treated as inherently moving forward, that is as the solution, have we even agreed upon the problem?" (Benjamin, 2019, p.180).

11. See Sasha Costanza-Chock's discussion of equitable design practices in Design Justice (2020). She argues that "[t]he participatory turn in technology design, or at least the idea that design teams cannot operate in isolation from end users, has become increasingly popular over time in many subfields of design theory and practice...[including] participatory design (PD), user-led innovation, user centered design (UCD), human-centered design (HCD), inclusive design, and codesign, among a growing list of terms and acronyms." (p.85) However, she also notes that many practitioners of these practices fail to "ask key questions about how to do design work in ways that truly respond to, are led by, and ultimately benefit the communities most targeted by intersectional structural inequality" (Costanza-Chock, 2020, p.85).

12. See Bryan Keith Alexander's "Embracing the Teachable Moment: The Black Gay Body in the Classroom as Embodied Text" in *Black Queer Studies* (2005).

sumes that students are often more engaged and successful in classes, and in their student role overall, when they occupy co-located learning spaces. The environment also assumes that physical proximity between students in face-to-face communities translates to closer, more positive, relationships with their classmates and with other students (West & Williams, 2017). These assumptions shape communal attitudes toward online, digital pedagogy and mark it as a more detached—less meaningful—form of teaching and learning.

Some digital learning spaces are organized almost exclusively around aspects of traditional face-to-face classrooms. Such spaces often offer similar structures for synchronous lecture and discussion time, in addition to assessments and feedback that are directly analogous in form and purpose. They may also maintain depersonalized policies guided by generalizable models within face-to-face environments. Here the digital medium has only a superficial effect on the pedagogical method. In this instance, the transition to online teaching represents enough change to justify pedagogical stasis.

Online learning is its own pedagogical form and not a derivative of face-to-face learning. These differences make room for new possibilities in teaching and learning practices that encourage us to rethink relationships between materials, reconceptualize the meaning and modality of interaction, create different pathways for community building, and adapt the learning process to foster transformative connections to/through work and ideas in each course. There are limitations to both approaches, but these are things to celebrate, not criticize. Critical instructional design invites us to consider exactly which contexts for managed contact we assume and replicate in this form of duplicative praxis—starting with ubiquitous learning management systems.

Ubiquity, in the case of learning management systems, is more a marker of successful management—of data, content, assessments, and connected tools, than it is of successful online placemaking or learning. Indeed, Sean Michael Morris asks "[w]hat if we were to theorize that the learning management system (LMS) is designed, not for learning or teaching, but for the gathering of data? And what if we were to further theorize that the gathering of data, as messaged and marketed through the LMS, has become conflated with teaching and learning?" (Morris, 2018, para. 1). Pushing this line of questioning further, we ask: what if we dissociated LMS data—or *managed* contact—from critical connection—or *meaningful* contact? How would this shape our vision of subjectivity, placemaking, or community-building when teaching and learning is online?

In *The Manifesto for Teaching Online*, Bayne et. al argue that "[t]eaching online reshapes its subjects in all senses of the word" and that this reshaping allows us to engage in digital place-making practices that are more accessible, inclusive, and—quite frankly—imaginative (Bayne et al., 2020, p.150-151). We see analogs to this in perennial conversations about infrastructure. For some, the term assumes a fundamental physicality that's expressed in roads, bridges, buildings, and rooms that make up our built environment. This assumption often supports a binaried view of physical and non-physical relations that shortchanges the non-fixed, iterative nature of digital pedagogy and its constitutive spaces. For digital forms of teaching and learning "[w]hat it means to be 'on the course' or to be 'at' college is never one thing; it is always multiple, enacted differently for every student almost at every moment" (Bayne et al., 2020, p.150-151). It does not hold the LMS as a fixed home base that simulates infrastructure(s) of physical proximity and relation. On the contrary, it assumes that place itself is a fluid co-construction that's differently important in digital learning (Bayne et al., 2020). This paradigm shift presents a challenge for models that are buoyed by forms of fixedness (ie: tradition, geographical prestige, physical borders/boundaries, etc.) where placemaking equals power. But what would it mean to frame this challenge as a chance for critical play through which transformative examples of placemaking and relation can surface? How might we imagine critical responses to traditional instructional design models through transgressive, hope-filled placemaking practices? How might this make learning more meaningful for teachers, students, and critical instructional designers? Our DIGC work speaks to how we have responded to these questions and offers a case study for digital teaching that supports meaningful learning.

Collaboration: DIGC in Context

How do we build critical instructional design practice into the infrastructural reality of an educational institution in the US?

The University of Pennsylvania School of Liberal and Professional Studies (PLPS) launched the Bachelor of Applied Arts and Sciences (BAAS) degree in the fall of 2019 as a model for emergent approaches to higher education. As Vice Dean Nora Lewis notes "[t]he goal of this new platform is to make an Arts and Sciences education more accessible, flexible, and affordable for working adults." This includes redefining "...who can get an Ivy League education by making it accessible to anyone who demonstrates the ambition and potential to earn it, without sacrificing the quality of the education offered" (Penn LPS Online, 2014). The degree program was designed to connect liberal arts education with professional and career goals through—mostly asynchronous—courses that followed a condensed eight-week schedule. The asynchronous nature of the courses made room for learners to choose the time and form of engagement that worked for them. It also shifted power to learners by giving them agency over the shape, pace, and place of their work.

These curricular design choices were developed through conversations with a faculty advisory board, administrative and technical support staff, and management executives from regional, national, and global employers. The program was also focused on affirming non-traditional student voices and needs in the development process. To that end, PLPSO conducted its own research with feedback from students and businesses who were ultimately interested in developing critical thinking, leadership, and communication skills. The degree program evolved accordingly, blending a liberal arts and sciences foundation with this applied focus. As part of this program, for-credit certificates were designed to provide learners with the opportunity to develop smaller credentials that could be earned either separately from the BAAS degree, or combined in pursuit of the full bachelor's degree.

The dual focus on an applied bachelor's degree program and a liberal arts education makes the Penn PLPSO program unique. This balance, integrating liberal arts as a necessary foundation for the curriculum, is not prevalent in other BAAS degree programs around the US and is specific to the Penn environment.[13] It's also an essential component of the DIGC certificate program. We could not have created DIGC without the support and active encouragement of LPS administration and faculty, who saw the potential for it to contribute an important element of the larger PLPSO program, including meeting foundational degree requirements.

The idea for the DIGC certificate program grew from an individual course focused on digital literacy and cultural change. Our early dis-

---

13. Most BAAS degree programs identify themselves in contrast with a liberal education founded in a traditional BA degree—that is, the BA is designed to prepare students with a liberal arts background for graduate education. They contrast the BAAS as a more vocationally-focused program that requires substantial academic education in "core academic [liberal arts] areas". (College Atlas, 2014)

cussions explored building a suite of courses within the BAAS that extended concepts of digital culture to discussions of technology from social sciences and humanities perspectives. The first step towards this involved developing a foundational course that invited students to explore digital literacy as a concept and practice that enabled them to shape present and future change. It picked up on humanist conversations about algorithmic bias, digital redlining, critical consumption, emergent strategies, and other concepts that worked to broaden the ways we think about, use, and repurpose technology.

After designing the first digital literacy course, we chose to complement it by building on the skills that students would need to understand the context of digital tools, and the culture changes that accompanied them. Rather than focusing on the theoretical and technological shifts themselves, the second course centered case studies that highlighted the different ways technological shifts affect people and their societies. Other course materials also supported this approach, including guest lecturers from folks who were invited to discuss the real world experiences of those technological shifts. The implications of these changes on the needs of contemporary digital culture were ever present in the course's historical narrative. The course unpacks how patterns of technological change from the past are reflected in the present, and speaks to the ways that students' own experiences color and shape their understanding of contemporary digital life. Like the digital literacy course before it, the design encourages learners to build community and relationships outside of the learning management system, disrupting norms of traditional course engagement.

Our conversation around a suite of classes developed fairly intuitively from there, as our interests also met a need in the structure of the growing degree program. In our proposal, we laid out our key concerns: to help students become critical consumers, designers, and creators (Colmon & Krieger, 2020). With these concerns as a foundation, we laid out visions for technological fluency that could create pathways to digital citizenship—and forms of critical liberation—that helped students shape change in the larger world. Since launch, learners have come with a range of needs—including folks who want to meet career and/or personal goals by completing an undergraduate degree, folks who seek particular credentials to advance their personal and professional interests, and folks who simply want to take classes on specific topics in digital culture.

What does it mean to be a teacher, and what is meaningful learning?

This question gets at the heart of our work and the work of critical instructional design which both seek to support meaningful learning. We see it as something that includes processes of co-creation that privileges interdependence between students and instructors. In other words, it's not something that an educator can define or decide in a vacuum. It requires meaningful relationships. This assumes trust that's affirmed through dialogue about course materials, the methods/means of engagement in courses, and what it means to teach and care for each other. After all, teaching is, perhaps first and foremost, communal care work that demands hope for the possible in pedagogical spaces—which affirm interdependent practices that support critical reflexivity and change.

Critical Instructional Design as Praxis

When we sat down to plan the DIGC Certificate, we were highly aware of:

- Our mutual concern for digital literacy, and the intersections between humanities and social science perspectives. We felt strongly that students would benefit from a hands-on, engaged, contextual approach to thinking about and using technology.
- The benefit of inviting learners to work with code. We chose to include learning fundamental concepts for at least one programming language for digital culture, which at the time of this writing is Python.
- The need to make room for students to apply work to their lives in ways that encourage agency inside and outside of course spaces..
- We developed a set of guiding questions for the certificate and for each course, which stems from a problem and question informed approach, including:
  - How does culture affect our understanding of a technology and how we use it?
  - What are our ethical responsibilities as consumer-citizens in using (digital) technologies?
  - How can we locate queer space(s) and time(s) within work

the embraces liberatory imagination?

- How does data influence what and for whom we code?
- What does it mean to study within a mutualistic community that challenges the ways we imagine sociocultural change?
- How might we explore the intersecting roles of race, class, gender, and sexuality in technology and culture to develop practices for engaging in equitable, inclusive, work across projects?

We quickly found that this inquiry-based framework fit the concerns we have for digital and cultural literacy and their related social and cultural contexts. They also give learners room (ala Freire) to take charge of their own learning practices and to own their own learning process.

As we developed the first two courses in the DIGC certificate, both of which are foundational to the certificate program, we examined the ways we wanted to build community for the classes. Because the PLP-SO program is primarily online-asynchronous, one key issue we had to explore was using and developing effective asynchronous discussion spaces. We wanted to present students with spaces for low-stakes communication that did not carry the burden of graded activities. In the case of the first course, Digital Literacy and Cultural Change (DLCC), these spaces outside the learning management system (LMS) were more intimate in that students were more willing to share information about themselves. They also found ways to make informal connections between course materials and their interests outside the course. Our choice to use workspaces outside of the LMS gives students room to collaborate organically, in ways that reflect their own growing ownership of their learning process and community building. Students share: files, links to artifacts they've created, projects that are in-process for the course, and poetry. They also share things that they would otherwise save for less formal social contexts, including memes, emojis, short videos, and playlists created or curated for their own use. We chose to use Slack, Coggle, Google Docs, and a host of other tools outside the LMS to build a non-hierarchical relationship across our courses. These choices were specific to the circumstance of the development of each course and to the goals for mutualistic collaboration. With both courses, the extra-LMS setting encourages flexible and non-intimidating space for students to critically evaluate their own ideas, in addition to societal assumptions, about digital culture and technologies. Since we did this in different ways in each of the courses we designed and taught, we decided to write about them in our own voices, sharing the practices we developed in each of these courses.

What does it mean to model being advocates for learning?

DIGC 120: Written from Clay's perspective

DIGC 120 began as it ended: with a poetry exercise. Some students were initially put off by this. Poetry wasn't exactly what folks expected from a foundational course focused on digital technology; but it offered a form of careful literacy that allowed us all to advocate for different—more sensitive—kinds of communication in our shared learning space. I paid close attention to democratic ownership when designing this and other activities in the course. Each of us drafted project prompts and peer-review guidelines together–as acts of creative digital making. We were also honest about the risks associated with such openness and vulnerability in the course. To support this, I included a note on sharing, risk, and vulnerability in the first week which read:

This course will require us all to grapple with concepts and realities that may make us uncomfortable at times. It'll also push us to share and reflect in a variety of ways. I've mentioned in the syllabus that sharing involves risks. Please know that my goal is to co-create a safe, inviting space in which all participants feel valued and heard. Still, some of us may decide that the risks are too high; and that's OK. Your assessment and feedback will not be contingent on how personal or transparent you are in this collaborative course space.

The teaching team holds power in this space, no matter how democratic we wish to make it. These few words are our way of acknowledging the imbalance that's built into this and other systems. We'll have a chance to talk about some aspects of these systems explicitly. We'll also try to mitigate various forms of imbalance as best we can, when we can. Although we've said it before, it's worth repeating: please share only what you're comfortable with sharing in the course.

We also wrote "notes to future students" that remixed poetic language and spoke to the generative aspects of vulnerability, as the course concluded.

In addition to poetry, I purposefully designed adaptable elements into the course structure that responded to co-created materials—like our community agreement which we authored, asynchronously, in a collaborative document outside the LMS. This initial activity encouraged everyone in the course to reflect on and share our individual responses to questions like:

- What kind of course will not work for us? What guidelines might we put in place to avoid replicating such a course for present and future learners?
- What shared expectations are important to us, and how might we hold ourselves accountable to them?
- What do we need to feel safe/engaged/seen/heard in our learning environment, and what would help make that possible?
- What brings us joy, and how might we incorporate that into the work we'll do together?

After sharing their individual perspectives, students joined small groups to synthesize the document into a collection of policies—which included gifs, poetry, song clips, and other components—that ultimately shaped presence and critical connection throughout the course. This established an infrastructure for care and mutual accountability early on; it also affirmed our intentions for knowledge-building. Instead of centering stagnant parameters that encouraged model responses, the course reflected our collective growth through contextualized work that we created as individuals and as a community. This form of agency was messy at times, even as it laid the groundwork for deep engagement with digital citizenship and non-traditional placemaking—on a small scale.

With the expectations of communal placemaking established, we continued work in other knowledge-building environments outside the LMS, including in mindmaps, informal discussion channels, audio-editing platforms, and low-stakes study spaces that were all driven by questions we each hoped to answer before the course ended. The question I asked invited us all to sit with how we might imagine and shape change via digital means when our current choices seem so overwhelming and limited. I reviewed and reshaped the contours of this and other questions each week alongside students. We talked about serendipitous finds, frustrations, and persistent fears in Slack;

we connected over music, soundscapes, and co-curated playlists in Spotify; and we worked through worldbuilding, capitalism, and commodified hope in SoundCloud. But the tools weren't the point; creating compelling environments for meaningful learning was. Lectures were narrative podcasts with unit-driven story arcs and interactive transcripts. Assessments were open-ended opportunities for serious critical play (that included Octavia E. Butler's science fiction and Adrienne Maree Brown's emergent strategies for shaping change). Outcomes were personal, adaptable, and interdependent (in response to Tressie McMillan Cottom's practice of "thick" revision). The course privileged design paths that eschewed fixed expectations and that encouraged us all to be sensitive advocates for learning.

DIGC 160: Written from Meryl's perspective

In building DIGC 160, a key concern for the course topic was to present a model for critical investigation of assumed knowledge. The goal was to balance giving a structure for student exploration of a new way of conceptualizing (in this case) how technological innovation takes place and giving agency to students in both their own learning and in supporting the learning across the course. Like my counterpart, I chose to use Slack as the primary mechanism for course activities and discussions because of its lack of hierarchical infrastructure—all participants in the workspace have the same capabilities to post, share, and engage with everyone else in that space. Because my course topic and approach was more of a social science focus than a humanities-based one, I had to use this kind of space differently.

Being aware of the need for clear and focused spaces for engagement, I minimized the number of channels where students were asked to share, and (with one exception) maintained consistency throughout the course in terms of creating public and small group spaces. Users in Slack can send Direct Messages (DMs) to any participant in the workspace, and can create ongoing conversations with a small group that is private from the rest of the workspace participants. I created channels for small groups and assigned students to specific groups to give them small (3-6 member) working groups they could build working relationships with, share knowledge, co-create and collaborate on a group project, and give and get constructive feedback on individual-level course assignments that were also accessible to the instructional team. Students also completed individual work within the course's LMS, where they could develop their own ideas with feedback from the instructional team. Criteria for evaluating these reflection submissions were processual—they were evaluated on the basis of instructions that guided them through growing their knowledge from whatever point they began from. Students in the class came from all parts of the US as well as two other continents, so norms of communication and learning, as well as personal experience, were highly varied across the class as a community. Students also ranged widely in age, which led to a wide range of perspectives being shared. Students also took this class for a range of reasons, ranging from the transactional[14] to different degrees of personal interests.[15]

In building the course infrastructure within the LMS, a similar set of concerns attended to the design of each week's content. Thinking of the LMS as a virtual classroom, where students and instructional team members all need to know how to access tools and resources, modules were designed with a consistent aesthetic, and to be accessible in as many ways as possible. For example, course lectures were recorded as podcast episodes of 6-11 minutes each, with transcripts of the episodes and relevant sources posted below each lecture. Students had the option of listening to lectures through the LMS on a computer, tablet, or smartphone, and also had links to listen through SoundCloud, should they prefer or be better suited to that experience.

Students were asked to process readings and course content through a reflection process that was consistent across the weeks of the course with instructions that were transparent—there was no assumption that the instructor role was the only one to introduce concepts to the class community. Indeed, the community of practice within the Slack space encouraged students to take on the role of being mutually accountable through regular group activities and discussion. Like in any learning process, this peer accountability process was sometimes uneven, but at the end of the day it created an environment that one student, in their evaluation of the course noted: "The Slack community was buzzing. This class felt really alive…" This vibrancy was fostered and supported by both members of the instructional team through regular and consistent feedback, and this modeling was recognized by the student members of the community.

---

14. DIGC 160 meets two different degree requirements, including as a Gateway course, and to meet the qualitative historical foundations requirement for the BAAS degree.

15. Students in DIGC 160 filled out a Student Background Survey, which asked them both about their reasons for taking the course, as well as their goals for how they hoped to use what they learned in it. 23 of the 24 students who completed the course completed the survey, giving a holistic view of the goals and interests of the participants.

I focused my practices on the framing of student-provided materials, sometimes reframing content to help students practice exercising their critical thinking muscles through the lens of the course structure. An example of this guidance was a comment I made on a student-shared news article and TedTalk on the topic of artificial intelligence and algorithms. These were both written/presented by an AI expert who was clearly representing a coding perspective that assumed it was unbiased and the student shared it as an example of this; one of the key concepts we were exploring in the course was uncovering the extent to which this is a logical fallacy, and the extent to which most people are unaware of such biases:

> Just putting on my professor hat here to make a comment on the approach of this article—it ignores that human bias is part of all incarnations of AI and algorithms. There was no superhuman who created the first one without any biases, any more than the first person to develop a marketable microphone didn't build in a preference for particular sound frequencies that matched what sounded good to their own ears. There is no such thing as an unbiased human. What there **can** be are humans who are aware of most of their biases and build checks and balances into the systems they develop to account for them." (Dr. Meryl, Slack #random channel comment, December 9, 2021)

At the end of the day, this example demonstrates the ways we met the goals of the class: to have students problematize and unpack the role of the digital algorithm as a fundamental building block of contemporary digital culture—they need a space in which to do it where they are free to be and engage as themselves, much as they do in DLCC. Future courses in the certificate program build on this foundation and are in the design process at the time of this publication.

Synthesis: What Would it Mean to Teach from Our Sensitive Edges?

Many of us sublimate our sensitive edges in our teaching. These include the places where we feel our most human and vulnerable. As Alexis Pauline Gumbs writes, our sensitive edges "enable us to live more porously, more mindful of the infinite changeability of our context, more open to each other and our own needs" (Gumbs, 2020, p.61). They also allow us to remain open to the generative messiness that supports human-centered learning. This openness makes room for us to practice self-governance which, in turn, allows us to connect with our desire to share, personalize, and change our teaching and ourselves. But it's not sustainable without creating a pedagogical infrastructure that affirms vulnerability and care as foundational to design decisions. This often involves embracing the things that make us unique as educators and learners—outside of flattening models—even when it challenges "the plan."

As a takeaway, at a theoretical level, it is important to reiterate how critical instructional design practices intersect with and disrupt normative models of education:

- **Power differences** are normative models that assume that instructors are the purveyors of knowledge, and that they are the funnels through which knowledge is passed to students
- **The asynchronous approach to learning** gives students agency in the sense that students have more control over their own learning processes
- **Disrupting normative models: Synchronous engagement in an otherwise asynchronous course** Synchronous engagement necessarily takes on a different, more democratic form. Many students who lack a framework for learning—either on a particular topic or are simply removed from the learning context—benefit from a structured, but open, space to directly engage with instructors and peers.

Our efforts affirm critical instructional design as an andragogical method that supports intentional placemaking practices. Class communities are co-constructed in digital spaces that work to ameliorate many of the barriers which exist in elite higher education and are relevant to any critically-designed learning space. In addition, the design of the BAAS supports the adoption of accelerated courses which similarly meets a well-identified need for learners to build the kinds of learning spaces that support their own needs and goals. Ultimately, this andragogical approach can benefit from more critical attention—particularly within the context of a post COVID-19 world which demands we ask different questions about access.

It's hard to overstate the pandemic's role in highlighting the ways critical instructional design methods intersect with the evolving landscape of higher education. While asynchronous online learning was a growing arena for higher education prior to COVID-19, the pandemic accelerated the exploration of new approaches to education while simultaneously foregrounding the need for adult learners to connect more intentionally with lifelong learning as a lens, rather than a stage intended for traditional age populations. This challenged many of us to think more critically about teaching and learning in digital spaces and to build from our sensitive edges through critical instructional design that prepares us for futures that demand change.

References

Alexander, B. K. (2005). Embracing the teachable moment: The black gay body in the classroom as embodied text. In S. P. Holland, M. G. Henderson, C. J. Cohen & E. P. Johnson (Eds.), *Black queer studies: A critical anthology* (pp. 249-265). Duke University Press.

Arnston, P. (1989). Improving citizens' health competencies. *Health Communication*, 1(1), 29–34. https://doi.org/10.1207/s15327027hc0101_4

Bateson, G. (1987). *Steps to an ecology of mind*. Jason Aronson Inc.

Bayne, S., Evans, P., Ewins, R., Knox, J. & Lamb, J. (2020). *The manifesto for teaching online*. https://doi.org/10.7551/mitpress/11840.003.0010

Benjamin, R. (2019). *Race after technology: Abolitionist tools for the new Jim Code*. Polity.

Colmon, C. & Krieger, J. M. (2020). *Penn LPS online certificate proposal: Digital strategies and culture*. Unpublished.

College Atlas. (2014). *Bachelor of applied arts and science (BAAS)*. https://www.collegeatlas.org/bachelor-of-applied-arts-and-science.html

Costanza-Chock, S. (2020). *Design justice*. MIT Press. https://doi.org/10.7551/mitpress/12255.001.0001

Davidson, J. W. (2021). Beyond trigger warnings: Toward a trauma—informed andragogy for the graduate theological classroom. *Teaching Theology & Religion*, 24(1), 4–16. https://doi.org/10.1111/teth.12574

Freire, A. M. A. & Macedo, D. (2021). *The paulo freire reader* (2021 edition). Continuum.

Gannon, K. M. (2020). *Radical hope: A teaching manifesto*. West Virginia University Press.

Gumbs, A. P. (2020). *Undrowned: Black feminist lessons from marine mammals*. AK Press.

hooks, bell. (1994). *Teaching to transgress: Education as the practice of freedom.* Routledge.

Knowles, M. (1970). *The modern practice of adult education: From pedagogy to andragogy.* Cambridge Book Company.

Morris, S. M. (2018). Reading the LMS against the backdrop of critical pedagogy. *Sean Michael Morris.* http://www.seanmichaelmorris.com/reading-the-lms-against-the-backdrop-of-critical-pedagogy/

Penn LPS Online. (2018). *What is a bachelor of applied arts and sciences degree?* https://lpsonline.sas.upenn.edu/features/what-bachelor-applied-arts-and-sciences-degree

Phillips, R. (2016). Organize your own temporality. In A. Romero (Ed.), *Organize your own: The politics and poetics of self-determination movements.* Soberscove Press.

SAMHSA. (2015). Trauma-Informed Care in Behavioral Health Services. https://store.samhsa.gov/sites/default/files/d7/priv/sma15-4420.pdf

Wartenweiler, T. (2017). Thinking about teaching: Does a student teacher impact the reflective practices of a cooperating teacher? *The Online Journal of New Horizons in Education, 7*(2), 96–106.

West, R. E. & Williams, G. S. (2017). "I don't think that word means what you think it means": A proposed framework for defining learning communities. *Educational Technology Research and Development, 65*(6), 1569–1582. https://doi.org/10.1007/s11423-017-9535-0

Militaristic Origins, Power, and Faux-Neutrality
A graduate student's interrogation of traditional instructional design

Hannah Hounsell

This summer, I had the most meta experience of my life. As a capstone to my instructional design graduate certificate, I was involved in designing instruction for Design Forward (https://bit.ly/CoLabDF), a program about critical instructional design. Oh, and in order to design the curriculum, I went back and, bit by bit, class by class, I critiqued the traditional instructional design training I received through my instructional design graduate certificate (as a fun activity, count how many times I say "instructional design" in this piece). When it came time to sit down and write about my experience, I floundered at all the meta. What piece of the meta was I going to reflect on? After being steeped in so much meta for the past three months, was it even possible to unravel it for an outside audience? And then, of course, as a university staff member and current graduate student, my age-old worries about what expertise I could possibly have in this area resurfaced. What follows is my attempts to resolve these tensions while also providing a perspective that furthers the field of instructional design... No pressure.

Unraveling the Meta

Let's start with some context. When my friend and mentor Robin DeRosa opened the Open Learning & Teaching Collaborative (https://bit.ly/OpenCoLab), AKA the CoLab, at Plymouth State University as a hub for teaching and learning praxis and community-driven academic professional development and I became involved in its inaugural days, I knew next to nothing about instructional design. Through my limited interactions with instructional designers when I was still a Program Support Assistant, I gathered that these were university staff members who helped faculty with instructional problems, particularly related to academic technology. Through my work in the CoLab, this understanding was questioned, unpacked, unpinned, expanded, and is still being defined and redefined—much like everything else I learn by working at the CoLab. When I started a graduate certificate in instructional design at the university in which I worked, it became clear to me that the instructional design we were doing in the CoLab was nothing like the instructional design I was learning about in my online classrooms.

So what's it like to work in an office where we frequently talk about turning education on its head by encouraging faculty to "ungrade" (https://bit.ly/IntroUngrading) their students, leave behind deadlines, cultivate classroom environments that are adaptable and embrace possibility, use open educational resources, etc.; and then, at night, I turn around and I complete my graduate assignments about ADDIE, Agile, the Understanding by Design Framework, and rubric creation? At first, it was nothing but frustrating. I felt like completing these assignments wasn't advancing my expertise beyond getting me the degree I needed to get into the classroom as an instructor. It felt like a repeat of my four years of undergraduate education courses (I pursued teacher certification), but with even more focus on structures and systems. Some of it felt directly opposed to the values, goals, and mission of the CoLab, a place where we encourage design based on care, accessibility, flexibility, and personal and public contexts. For example, in almost every grad course I took centered around curriculum, I was tasked with creating traditional, criterion-based rubrics for my units, without engaging in the active conversation in fields of pedagogy around the detriments (https://bit.ly/TroubleRubrics) of standardized assessment tools like rubrics. I felt ashamed, or like I was wasting time in a program that was antithetical to the values that were near to my heart.

However, when my capstone experience coincided with my colleague Martha Burtis and Jesse Stommel's piece on *Hybrid Pedagogy*, "Counter-friction to Stop the Machine: The Endgame for Instructional Design," and Martha's idea to develop an "emergent exploration of critical instructional design" through the CoLab's Design Forward program, I started thinking about how my frustration might be channeled into something productive. I was unintentionally engaging in critical reflection of the traditional instructional design curriculum that was guiding my graduate certificate. My experiences with a more critical pedagogy in the CoLab and my experiences learning about, writing about, and utilizing traditional instructional design were coming together. At their intersection was the development and articulation of my own approach to critical instructional design.

Speaking of meta, In 2016, as part of "MOOC MOOC: Instructional Design," a meta-MOOC designed for interrogating teaching and learning development, Sean Michael Morris wrote:

> Our corroboration of techniques for managing learning... is complicity, not agency. In many cases we have surrendered our agency for systems, theories, "best practices", that work smoothly, that take the effort of agency off our shoulders... There must be another way. And if there is, **we will discover it through critical reflection on the systems for managing learning which have proliferated** [emphasis added]... It will not come necessarily from knowing etymologies, nor from archaeologies, nor from our learnéd experts on learning (Vygotsky, Freire, Skinner, the whole lot of them) **unless they help us recognize the path we've tread that we must now transgress** [emphasis added]. (Morris, 2016)

Morris' reflections on the purpose that systems, theories, and practices serve in unearthing the well-tread teaching and learning trail have allowed me to find meaning in my traditional training. I approached Martha with the proposition that I help her with Design Forward by reviewing my traditional instructional design classes and providing specific critical reflections that could inform the Design Forward curriculum. My hope was that, in order for participants to, in Morris' words, transgress the ID path, I could help illuminate the traditional path we've tread.

Traditional ID is Based in Contexts that Need to be Interrogated

In my Introduction to Instructional Design course, we learned about the history of instructional design using Reiser's "A History of Instructional Design and Technology" (2001). Instructional design has origins in World War II where psychologists and educators were recruited to design military training to prepare millions of drafted soldiers. These psychologists went on to research, publish on, and develop the field of instructional design. Modern instructional design also has corporate influences. Throughout the 70s and beyond, the desire for increased productivity and business results influenced instructional design, especially the growth of a systems approach to tackle specific revenue goals. The use of technology and the need for training around programs and software in the corporate world precipitated growth in the instructional design profession.

I wonder how many military and corporate influences we can still see in our modern instructional design utilized outside of these contexts? For example, the air force was instrumental in the development of aptitude testing to determine whether or not potential airmen had the skills necessary for flight training. Aptitude testing is utilized in higher education from determining student acceptance to placing students in math, science, and language classes. An article out of eLearning Industry, "Instructional Design: A Military Perspective," by Instructional Technologist for Helmerich & Payne Int'l Drilling Company, David Mallette, provides an eye-opening perspective on ID that I think will sound really familiar to anyone who has ever had formal ID training. Forgive the long quote, but I think the whole thing is necessary to see the hold that its military origins still has on instructional design:

> Given a clearly stated performance problem, the instructional systems technologist will provide a firing solution consisting of a **performance-based, criterion referenced objective and the lowest cost, highest efficiency medium** [emphasis added] to deliver it in 4 hours or less. I use the term "firing solution" deliberately because there is a direct historic military analogy. The Allied bombing campaigns of WWII used what is termed "saturation bombing" strategy. Due to the lack of guidance and aiming tools, hundreds of bombs would be dropped in the hope at least a few would hit the target. In spite of the cost in lives and resources, it was the best they could do and deemed necessary. Today, **the military uses precision guided weapons carried or propelled by precisely designed delivery systems to hit the target precisely. Training professionals should too** [emphasis added]. The performance-based, criterion-referenced objective delivered using the lowest cost, highest efficiency medium is our tool (Mallette, 2012).

Commonly used instructional design practices that mirror the most cost-effective method for killing people is a grim picture. At the very least, this passage demonstrates the military and corporate origins and influences on traditional instructional design's obsession with performance outcomes, criteria and objectives, and low-cost/high-efficiency solutions. These terms are familiar to any instructional designer or instructor. Tie learning to outcomes. Outcomes to assessment. Do more with less. Create workers/soldiers/students ready to perform and produce. Should practitioners in the field of education base their teaching and learning on systems designed to churn out capable soldiers and increase profit margins? These origins and the systems they engendered can't go unquestioned. Instructional designers need to consider the purpose of the education they design and what structures and systems can support those purposes. Or better yet, can systems alone support our purpose?

Traditional ID Thrives on and Perpetuates Hierarchies

From the very first instructional design class I took, I was struck by these thoughts: Why aren't we talking about pedagogy? Why aren't we talking about students? Why aren't we talking about humans? I think it's because, through the corporate/military lens of traditional instructional design, people don't matter as much as outcomes. Winning the war. Increasing productivity. Increasing the bottom line.

Another perspective is that a "people agnostic" approach in the instructional design profession helps establish boundaries and barriers between the field of teaching and the field of instructional design. From the very beginning of my graduate coursework for Curriculum & Instruction and Instructional Design, there was a differentiation established between "instructional design" and "actual teaching." This differentiation precipitates a hierarchy between those who design instruction and those who teach. In the instructional design profession, subject matter expert (SME) is the name given to the teacher that works with the instructional designer. Even in my intro texts, the power dynamics between these two roles were obvious. Instructional designers are experts in the many design structures and frameworks—ADDIE, Agile, UbD, the 9 Events of Instruction, Bloom's Taxonomy, etc. Instructional designers support the integration of technology or help design in online spaces. Instructional designers aren't teachers. In *ISD From the Ground Up: A No-Nonsense Approach to Instructional Design*, one of my intro texts, author Chuck Hodell writes this bit of advice: "Work as an instructional designer—never a subject matter expert—while designing. In the same way that a designer does not want to be challenged on design expertise, **a designer should never challenge a SME on subject matter expertise.** Fair is fair. It is often best to state your role early on—**that any questioning of the content is meant for clarification** [emphasis added]" (2016, Working Effectively With Subject Matter Experts chapter). We don't even have to read between the lines here. The message is clear: instructional designers should stay in their lane. Questioning is means for clarification and should not be misconstrued as critical. Do not challenge the almighty SME.

And by the way, the instructional designers that I have met in my time within institutions of higher education have all held staff roles. I don't think that's a coincidence when we are talking about power dynamics between instructional designers and SMEs. University staff are told to stay in their lanes and "act like staff" in a thousand tiny (https://bit.ly/

StaffMicro) and big (https://bit.ly/StaffNoSay) ways every day of their professional lives.

It's hard not to continuously quote Morris, so I'm going to do it again and I'm not even going to apologize for it. In 2018, Morris wrote the article "Instructional Designers are Teachers" and discussed the ways that traditional instructional design programs weaponize their structure-obsessed curricula to maintain power dynamics within ID.

> Skills-based and outcomes-based, positivist-inclined, Bloomsy instructional design training programs, often disguised as advanced degrees or certificate programs in online teaching, **have perpetuated the notion that instructional designers are computational, technicist, and mechanistic in their work** [emphasis added]. Most instructional designers are taught that their work should consist of aligning outcomes to assessments, assessments to assignments, assignments to content. That they should be masters of Screencast-o-Matic and Powerpoint and Voicethread rather than teachers. Courses are not compositions for these instructional designers, but rather blocks fit together and packaged, tied round with a rubric for a bow (Morris, 2018).

And these power dynamics play out in other ways besides the instructional designer/ SME dichotomy. Traditional instructional design's obsession with structures and frameworks leaves very little room for student choice and agency, emergence, and flexibility. In the always spot-on words from Maha Bali, "…by making concrete, discrete decisions about what will be learned in advance of learning encounters it ignores the differences, interests, capabilities of both teachers and learners" (2014). This approach supports designers and teachers as the undeniable authority over learning.

I don't think that my ID teachers were nefariously rubbing their hands behind their online courses and cackling about their evil plan to oppress future instructional designers via torture by Bloom's taxonomy. But I think the history and trajectory of instructional design has gone unexamined by practitioners. This is a self-perpetuating issue because a profession built around rejecting inquisition is inclined to replicate itself without reinvention.

Traditional ID Tries to Convince You that it is Neutral

To return to my earlier statement, my introduction to professional instructional design did not include conversations about people. Ok, yes, let me be clear, we talked about how people learn, remember, engage, motivate, and how to design learning experiences that use evidence-based best practices to encourage learning. So in that way, we talked about people. But whatever conversations we might have had around learning design as a tool for perpetuating or breaking systems of oppression or discussing design in the context of inclusion, accessibility, affordability, and equity, we replaced them with conversations and assignments around systems. I think I've made it clear by now: traditional instructional design is obsessed with structures. It's because structures are safe. Structures feel neutral, nonpolitical, objective. Rubrics feel neutral, nonpolitical, objective. Learning aligned with goals and goals aligned with assessment feels neutral.

Another quote from the intro text that Hodel wrote illustrates this sentiment about ID perfectly: "The neutrality of the ISD [Instructional Systems Design] process is vital to its effectiveness as a system. It must not contain any inherent bias or preconceived notions about any aspect of a particular design process. This is vital when an instructional designer is making important decisions about various aspects of a course design, including implementation choices and other noncontent-related areas" (Hodel, 2016). Although Hodel specifically speaks of instructional designers having no biases around which systems of design are better, whether intended or not, this quote presents instructional design as a practice that is somehow free of bias or prejudice. As if design isn't often weaponized (https://bit.ly/WeaponizedDesign) and exclusionary (https://bit.ly/ExclusionaryDesign).

It's easier to perpetuate the myth that instructional design is neutral because we have established a clear hierarchy between instructional designers and teachers. It's easier to think of back-end professionals who fit instructional activities into tidy, pre-determined structures and who align learning to outcomes, outcomes to assessment, assessment to content as neutral beings. I'm reminded of "personalized learning" fantasies (https://bit.ly/TechFantasies) which suggest learning and teaching can be automated, objective, free of human error... because how (https://bit.ly/BiasedTech) can something automated be biased?

Toward a Critical Instructional Design

I encourage all instructional designers who have been led to believe that they aren't teachers to resist because it's the first step in understanding that instructional design is not neutral. Teaching and learning are acts of liberation and thus designing teaching and learning is an act of liberation. Folks are empowered to transform an oppressive society through education. Privileged folks become aware of their own privileges. Systems of privilege become interrupted. This is the basis of Critical Pedagogy. In "Life in schools: An introduction to critical pedagogy in the foundations of Education," McLauren (2015) writes that we must push past the reductive understanding that the purpose of Critical Pedagogy is simply to empower and transform; we must ask ourselves who is transformed/ empowered by Critical Pedagogy and for what end (p. 9)? We promote critical thinking in our classrooms. We use classrooms as sites for rich, even controversial discussions, embracing and utilizing students' differing perspectives. However, we must ask ourselves, as educators, to what end is critical thinking, debate, and discussion? We don't build these skills in students because of what Bloom's Taxonomy or ADDIE tells us; we build these skills in students because they empower our students to enact political and social change. We can interrupt oppressive cycles with learning and teaching:

Sometimes it might be some new awareness or consciousness that we gain. Perhaps a friend from a different identity group shows us a different perspective, or we read a book that makes us think differently, or we enroll in a course that introduces new possibilities. We begin to see the big picture that groups all over the world as working on these same issues... once you know something, you can't not know it anymore, and knowing it eventually translates into action (Harro, 2018, p. 33).

Political neutrality isn't possible in the classroom because the very act of teaching our students is political. Our purpose is not to pump our students with information and train them to conform; our purpose is to teach them how to question, challenge, and change the world they live in. Instructional designers are engaging in a political act when they design learning.

Instructional designers are also engaging in a political act when they use systems that go uninterrogated. Neutrality supports, maintains, and strengthens the status quo. And the status quo is often designed to oppress. According to Kandel-Cisco's "Avoiding neutrality in climates of constraint: Moving from apathy to action" (2018), "Educators... exhibit complicity through their silence... passivity, while seemingly apolitical, is actually a political act in itself—an act that demonstrates satisfaction with the current state of affairs" (para. 10). Oppressive ideologies like white supremacy are intrinsic and pervasive in our systems and, because they are politically, historically, economically, and psychologically replicated, they often go unquestioned. I encourage instructional designers to resist a preoccupation with maintaining neutrality and using uninterrogated systems. Their systems grew out of the military. They were influenced by corporate and capitalistic needs. They are based on sociological and psychological studies fraught with their own history of bias. Some instructional design practices replicate oppressive systems. Amy Collier says it quite beautifully: "Design is not a neutral activity. We design through our own lenses, assumptions, politics, goals, beliefs about the world–through our own humanity... Our instructional designs–both digital and analog, implicit and explicit–embody what we believe about students, about education, about the goals of learning" (2017). So if Collier is right that design is not neutral, then it is biased and a critical instructional designer should pull that bias toward justice and learner empowerment.

Ending with an Anecdote

At the end of June, right when I was gearing up to present my mini-unit (https://bit.ly/tradIDslidesrevised) during Design Forward on the origins and evolution of instructional design, an interesting conversation took place in the CoLab's "Teaching & Learning" Microsoft Teams where all instructors in our institution are welcome to take part in informal conversation around pedagogy. A professor asked for some resources and advice about the psychology of deadlines. In particular, they wanted to know if there was research about deadlines improving performance and task completion. They noticed that a lot of students struggled with the completion of assignments and self-regulation when they relaxed their deadline requirements due to COVID... and a lot of students appreciated the flexibility. The professor wondered if there's a sweet spot between flexibility and strict deadlines where students felt like they had a healthy structure to support their success without making it punitive. Another professor posted a research paper as a response, "Cognitive performance is enhanced if one knows when the task will end" by Katzir, Aviv, and Liberman.

In a lot of ways, this article seemed contradictory to one of the values of Design Forward and critical instructional design—that structure dampens humanity, individuality, creativity, and possibility in a classroom. After reading my piece, it might seem like I think that critical instructional design calls us to pick up our pitchforks and torches and chant "down with all structure." But I think that structure and possibility can coexist. In everything we do, we are bound to structure in some ways. For the pilot of Design Forward, we were bound within the four weeks that the workshop was scheduled for. We bound ourselves to working inside a Google Document and utilizing a HyFlex modality. In traditional classes, we are bound by the beginning and end of a semester. We are bound by the need for students to submit all assignments before grade cutoffs. We function within these "beginnings" and "ends" because that's how human society works. But I think that critical instructional design asks us to interrogate design frameworks, practices, and origins instead of accepting them because "that's how we've always done it." It asks us to never sacrifice humanity for the efficiency of a pre-packaged process. And it asks us how we can work within the confines of immovable beginnings and ends, while also allowing human messiness intrinsic in learning and teaching to flourish.

References

Collier, A. (2017). Design and the beauty of palimpsests. *Digital Learning & Inquiry.* https://dlinq.middcreate.net/dlinq-news/design-and-the-beauty-of-palimpsests/

Bali, M. (May 3, 2014). Curriculum theory, outcomes/objectives, and throwing the pasta out with the pasta water. *Reflecting Allowed.* https://blog.mahabali.me/pedagogy/curriculum-theory-outcomesobjectives-and-throwing-the-pasta-out-with-the-pasta-water/.

Burtis, M. F. (2021) Design forward: An introduction. *Open Learning & Teaching Collaborative.* https://colab.plymouthcreate.net/2021/06/04/design-forward-an-introduction/

Harro, B. (2018). The cycle of socialization. In M. Adams, W. Blumenfeld, D. Catalano, K., DeJong, H. Hackman, L. Hopkins, B. Love, M. Peters, D. Shlasko, & X. Zúñiga (Eds), *Readings for diversity and social justice* (p. 427-34). Routledge.

Hodell, C. (2016). *ISD from the ground up: A no-nonsense approach to instructional design* (4th ed). Association for Talent Development. [Kindle Edition].

Kandel-Cisco, B. & Flessner, R. (2018). Avoiding neutrality in climates of constraint: Moving from apathy to action. *The Educational Forum,* 82(3), 290-302.

Mallette, D. (October 3, 2012). Instructional design: A military perspective. eLearning Industry. *eLearning Industry.* https://elearningindustry.com/instructional-design-a-military-perspective

McLaren, P. (2015). *Life in schools: An introduction to critical pedagogy in the foun-*

dations of education. Paradigm Publishers.

Morris, S. M. (April 12, 2018). Instructional designers are teachers. *Hybrid Pedagogy.* https://hybridpedagogy.org/instructional-designers-are-teachers/

— (2016). MMID: Toward a critical instructional design. *Digital Pedagogy Lab.* https://digitalpedagogylab.com/toward-a-critical-instructional-design/

R.A. Reiser. (2001). A history of instructional design and technology. *Educational Technology Research and Development*, 49:2, p. 57–67. http://faculty.mercer.edu/codone_s/tco665/2014/History_of_Instructional_Designtwo.pdf

Stommel, J., and Burtis, M. (2021). Counter-friction to stop the machine: The endgame for instructional design. *Hybrid Pedagogy.* https://hybridpedagogy.org/the-endgame-for-instructional-design/ [Republished in this collection]

Towards a Critical Instructional Design Framework

Katrina Wehr

As Costanza-Chock (2020) points out in the concluding chapter of Design Justice, "we urgently need more critical analysis in every design domain." What critical analysis might look like for instructional design, however, is complicated for practitioners who are rarely afforded final authority or power over the design itself. While existing scholarship in this area offers a variety of approaches and considerations, the literature does not point to a single ideal framework or list of steps for assessing instructional design in terms of inclusivity, power, equity, or justice. In order to get closer to critical practice of instructional design, we in the field must blaze the trail ourselves. In this chapter, I explore the ways power, process, and positionality influence course design from a perspective many instructional design professionals are familiar with: an inherited course, that is, one we did not design ourselves but are now tasked with maintaining and improving.

For practitioner-researchers like myself, this leaves room for exploration into how one might critically analyze an existing instructional design in an effort to develop a framework for designing more just learning experiences from the start of the process. The more we practice these types of design approaches, the more mature the processes will become (Costanza-Chock, 2020). To begin, I decided to analyze a course in my instructional design portfolio based on existing critical scholarship in order to practice interrogating these concepts as they play out in a course design.

The following work is an effort to create salient applications of themes identified through research and practice regarding design justice in instructional design contexts. In a 2020 article, Collier (2020) draws on the writings of various educators which implores us to approach critical analysis of designs by working to uncover "what's wrong" in order to take action towards counteracting marginalizing designs in education. This paper draws on critical scholarship in the learning sciences to guide the analysis and help other instructional designers focus their own critical analyses of existing course designs by providing examples of my thinking about these ideas in relation to specific elements of an online course. I also attempt to synthesize my process into a framework as a starting point for improvement with future critical analysis cycles. At the conclusion of this paper, I summarize my discoveries and reflect on the process and my role as an instructional designer.

Framing Concepts

Instructional designers are well positioned to influence the design of learning environments, and we should be "accountable for the social and political consequences" of our work (Barab et al., 2007, p. 296). I view my role in this critical analysis as similar to that of Barab et al. (2007). I aim to "position [myself] in a manner that will attune [me] to extant issues and highlight them among the community" (Barab et al., 2007, p. 281). The community for this work includes not only instructional designers on my team and instructors who teach this course, but anyone interested in critically analyzing existing instructional designs and how their discoveries are connected to, and could be changed by, the processes they follow to design instruction.

To begin the analysis, a review of critical learning sciences research was conducted to derive a framework to evaluate the instructional design of the focal course in this analysis. "Questions of power and ideology come to center stage in the decisions that educational designers make" (Barab et al., 2007, p. 291), and it is important that this framework is a flexible one that could potentially be utilized to conduct a similar analysis of other courses. This effort attempts to critically interrogate an existing design with the goal of uncovering areas for improvement, or, as Collier (2020) writes, to discover "what's wrong" with a design.

Taylor (2018) outlines some commitments to ethical teaching and research, such as the commitment to "foregrounding issues of historicity, race, power, and privilege in the curriculum I teach and/or design" (Gutiérrez & Vossoughi, 2010, in Taylor, 2018, p. 196). Taylor identifies some questions that guide this commitment, which I find particularly valuable to this work: "What voices are unaccounted for here? Why might that be? How can I/we fix that?" (Taylor, 2018, p. 196). While these questions of representation won't reveal every opportunity for improvement, beginning a critical analysis with this mindset will uncover details about the course design that can lead to other avenues of critique.

Barab et al. highlight power and its role in design. Indeed, power is conveyed, wielded, and distributed in different ways in a learning design. Esmonde (2016) discusses the relationship between power and artifacts, especially in a learning environment, and cites the example of curriculum standards and how their existence shapes the way K-12 teachers in the U.S. not only teach students, but how their performances are evaluated as well. Similarly, a learning design in an online environment in higher education generally includes predetermined learning objectives and disciplinary practices dictating how learners participate and develop their skills, or in other words, what types of practices and knowledge are considered valuable and worthy. Examining these arrangements of power and authority in a learning design is crucial to critical practice of instructional design.

Questions of identity and values are also essential to a productive critical instructional design effort. When considering identity and how it is relevant in learning, Vakil (2020) implores researchers to consider not only who students are in the present, but also who they have the potential to become as they learn. In practice, this means thinking about the diverse racial and gender identities taken up by learners and how those identities are reflected, or not, in an instructional design. Drawing on Wenger (1998), who stated, "learning is an experience of identity," Vakil argues that as learners' identities change, those shifts, "give rise to new ways of making meaning of and interacting with the world," as well as their relationship with the world (p. 92). But throughout this process of shifting identity, learners are also considering what their disciplinary learning means for their "possible futures" (Vakil, 2020, p. 93). As a result, learners may question the narratives surrounding political and ethical values of their chosen disciplines, such whose perspectives are considered expert lenses in the field, leading learners to "consider the kind of person one has to be, or become, in order to participate in the communities of practice explicitly or implicitly associated with particular forms or domains of knowledge" (Vakil, 2020, p. 93). This can lead to both positive and negative impacts on learners' disciplinary learning and their identities when those disciplinary values align or conflict, which makes ethical identity development and ethical values important commitments to this critical analysis.

To summarize, the extant literature points to three commitments, or for the purposes of this analysis, critical focuses, that create a map through this analysis. Taylor's (2018) questions of representation, Esmonde's (2016) ideas around power, and Vakil's (2020) concept of disciplinary values serve as themes for exploring how a course design could be improved for more just outcomes.

A Framework for Critical Instructional Design

Using these critical focuses to guide the analysis, I also turned to Vossoughi and Gutiérrez (2016) to identify specific units of analysis within the design. In their discussion of critical pedagogy, Vossoughi and Gutiérrez (2016) challenge researchers to consider the "how" of teaching rather than "what is to be taught" (p. 143). This switch in focus calls for critical analysis of how learning is organized, how practices situate power and ideology, how social relations are established, and "how tools expand or limit opportunities for development of critical thought" (p. 143). Based on my experience as an instructional designer of online learning experiences, these units of analysis can be neatly aligned to the main component parts of a learning design. The way learning is organized could include not only the linear progression of a course as laid out in the syllabus, but also the learning management system (LMS) and choices about how learning is presented in a digital space with regard to features enabled or disabled in a given design. When considering power and ideology embedded in practices, I look to examine the activities and projects learners are completing. Social relations between instructor and learners, and between learners, can also be analyzed based on the practices included in a learning design and can be observed by looking at discussion posts and announcements among other artifacts. Examining the course policies and procedures around topics like late submissions and test taking can also answer questions regarding power and values. Finally, the tools, such as software, LMS integrations, and others, that are utilized both by the instructor and learners play an important role in online learning experiences and are important to examine in a critical analysis. These units of analysis identified by Vossoughi and Gutiérrez serve as focal points for uncovering data related to the critical focuses identified previously. The table below organizes this framework and includes the types of course data I will analyze. "Units of analysis" has become "sites of analysis" to clarify that those design components are where one should look for evidence related to the critical focuses. The following analysis examines a foundational course that serves as an entry point into a completely asynchronous bachelor's degree program at a large research institution in the U.S. I used the master course space for the purposes of this work and investigated these areas of focus on my own in an effort to develop this framework, but in the future, this process could and should occur in partnership with other instructional designers and even faculty partners. Next, I will highlight specific examples from the focal course and expand on this framework, followed by suggestions for improvement or practices to consider replicating in other learning designs.

Critical Focuses	Sites of Analysis		
Identity/Disciplinary Values Who do learners become in this discipline? How do identities change	**Organization** content, syllabus, procedures, policies, reference materials, books, example work from previous learners	**Social Relations** learner/learner instructor/learner	
Representation Who is here? Who isn't here? Who should be here? **Power** Who matters? What matters? Who decides?	**Tools** LMS, third-party LMS tools, Studio, software	**Practices** learning goals, activities such as discussions & peer reviews, projects & assignments	

Table 1: Draft Critical Instructional Design Framework

Organization

While many design components of any course would fit the organization category of analysis, for the sake of brevity this chapter will be limited to 1-2 components from each category. Of all the components that organize learning in the course, I focus here on the content. Since content itself is a broad term, this paper considers reference materials such as books, instructor-provided documentation, and examples of work as primarily composing course content.

The focal course is based in the arts and is made up of five lessons. Each lesson is structured to include reading a selection of chapters from the required text, and a book that is reviewed as a critical reflection on the role of the discipline in everyday life. The lessons also contain other reference materials in formats such as audio podcasts and videos. In addition, each lesson contains some form of practice activity and culminates in the completion of three projects. An outline of the topics covered in the course is included below.

- Design Thinking – systems thinking, critical thinking, design process
- Visual and Interaction Design – semiotics, inclusive design, critical design, visual design, identity design
- Storytelling – structure, development, character
- Open Design
- Self Design

At first glance, the organization of learning as it is reflected in these materials answers some of the guiding questions about disciplinary values and learner identity. Learners are introduced to the idea of using discipline-based insights as a frame to understand complex problems and develop comprehensive, critical solutions. As an introductory course that ushers students into the discipline at this university's art school, the course topics prioritize key concepts alongside critical concepts relevant to the field, positioning these issues as foundational to the discipline. If they were not already, learners in this discipline will become practitioners who are at least made aware of the importance of these topics and will be expected to incorporate them into their practices within the course. A noteworthy observation about this course involves the perspectives from which these critical concepts are presented, a matter that will be addressed later.

Although I was not the instructional designer in charge of this course's initial development, I am aware of the processes and decision making that went into it as a member of the unit within which the development was supported and this is where the questions of power regarding organization of learning are answered. In my department, the instructor is the ultimate authority on what content is included in a course, and in what order, which has the tradeoff of leaving out the voices of learners and other instructors within the department. This arrangement creates an additional barrier for instructional designers and our ability to make positive changes toward more equitable and inclusive instruction. When control over the course rests so singularly in the hands of one individual, there is a risk of replicating that instructor's ways of knowing and being within a discipline, for better or worse. In the case of our focal course, the instructor chose to prioritize inclusive, critical discipline-relevant concepts alongside foundational disciplinary concepts, but not every individual instructor may have made that same choice. An area of improvement when it comes to power and authority in learning designs across the board would be to consider who the stakeholders are more broadly and include additional stakeholders in the process to ensure the voices of others are present in the course content. Additional stakeholders could be other faculty, a department chair, or even students.

Finally, representation in the organization of learning in the course can be analyzed by examining the authors who are cited in the course to learn who is present and who is missing. As part of an independent project, a past student kindly shared their data about representation in the course content which was collected, along with similar data from all the courses this student had taken so far in their program of study, for a project in another class. I verified these numbers and share them here with permission and gratitude to the student who chose not to be identified. Of the 59 scholars whose work appears in the course, 58 are white and of American or European descent, one is a Hispanic male, and seven are white women. To connect this oversight to an earlier point about power in the learning design process, bringing more voices into the process may have illuminated this lack of representation earlier in the process.

Social relations

Next, I will analyze how power and identity development are intertwined in the social relations supported in the learning design. Vossoughi and Gutiérrez don't offer a specific definition of social relations in their chapter, so I lean on the instructional design definition of social interaction in online courses which centers on interaction among classmates and instructors. By this definition, there are two primary social relations that are specific to this class, which are learner-to-learner interactions and learner-instructor interactions.

There are many opportunities for learner-instructor interaction in the course design, as instructors provide feedback and grades for every activity learners complete, and learners are invited to use the commenting feature in the LMS to communicate with instructors regarding submissions. However, there is only one formalized opportunity for peer-peer interaction where learners can offer each other feedback and interact. Peer feedback in the form of studio critique has its own dedicated section in the course policies and procedures which outlines how feedback should be given and received, thus putting up boundaries around how students interact with each other during these activities. Such guidance cites the "critique sandwich" method of sandwiching suggestions for improvement between one to two positive comments. The critique guidance also offers tips on how to make this process useful: "If someone gives you ambiguous feedback, this means that they can intuitively see a weakness but might not know why something isn't working. You should follow up with their comments with probing questions to better understand their perspective." Learners are graded on critique as participation, so it can be assumed to be a disciplinary value in the design discipline, and learners could anticipate becoming comfortable with the idea of critiquing their peers with advice and criticism in the spirit of improvement.

Critique is a commonplace teaching strategy, especially in arts-based disciplines, but the process of giving and receiving feedback in this way is a practice entrenched in a culture that is worth questioning. The "critique sandwich" structure in particular does not allow for learners to express their positionality when providing a response to classmates' work, but rather encourages students to make a value judgment on others' work as good or bad, a practice with ties to white supremacy culture (Okun, 1999). While boundaries and structure have a beneficial place in a foundational course, an approach to critique that encourages appreciation for classmates' efforts and recognizes there are a multitude of ways to provide substantive feedback could be a more productive method.

Critique sessions may be viewed as an opportunity for learners to demonstrate their identity development as an artist, and possibly distribute expertise among the class rather than looking at the instructor as the sole expert. Nowhere in the course is it explicitly stated that an instructor's feedback during critique carries more weight or should be treated differently from peers, but since the instructor is the one handing out grades, it isn't unreasonable to assume that instructor feedback may be perceived as non-negotiable. While the critique process may seem like a way for learners to practice being experts and hold some of the power within the course, ultimately the instructor likely maintains the most power in this activity as a result of setting the critique guidelines and also determining what an acceptable critique is. These social relations as designed are more related to identity development and practicing using disciplinary language and ways of interacting than they are designed to disrupt the power balance in the course.

Tools

The next focal point in this analysis is the tools utilized in the course learning design. Tools can both constrain and empower, and in an online learning experience, tools mediate learner's interactions with learning materials, classmates, and their instructor. Since tools are so ubiquitous in online learning and mediate nearly all aspects of student learning, analyzing the values, power, and opportunities for representation that tools do and do not provide is essential to this work. There are several software tools learners utilize to complete projects, including InVision App, Twine, and Adobe Spark, which are available for use without purchase. The use of publicly available tools in a foundations course could convey to learners that specialty software is not necessarily required to do real design work, and dispel any potential notions that using higher cost tools produces higher quality work. On the other hand, these tools are very specific in nature and may constrain the projects learners produce in ways that are counterproductive to learning about design.

In addition to the software students use to produce projects, there are tools that students utilize to access course materials, submit assignments, and interact. This course is unique in that the learning design consists of three systems that work together to provide different functions. The LMS (Canvas) serves as the course home base and grading center because it is supported by the university and tied directly to the student information system which allows for seamless access without requiring students to create multiple accounts and provides customizable options for learners to set up notifications that remind of due dates and alerts for instructor communications. Tied into this system is the virtual studio tool, which sits outside of Canvas specifically for critique of in-progress work. The virtual studio tool was developed to provide a visual-based discussion area for students to share many different formats of work, such as audio/video files, large images, animations, etc. Third, the course uses GitHub to manage updates to the content and deliver course pages in order to maintain consistency across concurrent offerings. The GitHub site is accessed when learners click on the link from the course Canvas site and is positioned as an electronic textbook.

A critical analysis of the Canvas LMS could be its own book, so for the purposes of this analysis I will focus on the virtual studio. The studio primarily supports the disciplinary values of design by prioritizing visuals as the objects of discussion. Learners typically photograph their work from various angles (for physical objects) or upload a selection of in-progress screen shots for digital work, and provide a short artist statement for their classmates to accompany the images. This foundational course emphasizes the design process, so learners are encouraged to document their work at various stages and share it. Learners who may previously have only presented finished work in past educational experiences would ideally become people who are comfortable with the idea of work in progress through this sharing activity, and understand that projects are iterative and open to criticism and improvement.

In terms of "what matters" in the studio, comments and visuals are the primary features of the tool. The only options learners can utilize when they click on a peer's work is to scroll through their images and add comments. By prioritizing the visuals and the comments submitted by learners, the studio reflects the values of looking at works in progress and considering feedback. As was mentioned earlier, the development of the virtual studio is another project that was primarily driven by one instructor, though more staff members were part of the process due to the complexity of skillsets required to complete this type of undertaking. Programmers, UX designers, and the instructional designer were all part of the team, but still student voices were absent from the development of this tool, even though it was designed specifically to support online learners' studio activity. When using a specially-designed tool like this, learners' activity could be constrained by the need to fit the submission model.

Practice

Finally, practice is the last unexplored site of analysis. When examining a learning design from an instructional design perspective, practices are considered to be the activities learners undertake that are guided by a set of goals. When it comes to the learning design of this class, components that I categorized as practices include the learning objectives, projects, and activities that learners participate in. While projects and activities may seem obvious, the inclusion of learning objectives as practices may initially be unclear, but I justify this choice because the learning objectives in any course ultimately guide the assignments, exams, projects and other activities learners engage in. Experienced instructional designers understand the learning objectives of a course to be an explicit statement of expectations for both the instructor and the learners. In an ideal world, the objectives are the guide posts that drive the learning experience, thus they are part of practices.

In addition to guiding learners' progression, the learning objectives should also illuminate the disciplinary values of any course. For example, the focal course states the following objective: "Implement new ideas and develop a diverse array of options for problem solving in response to critical review and the iterative design process for improving work." Learners might interpret from this objective what values are prioritized in the discipline and the course, which align with the values of iterative work, sharing, and openness to critique that I have uncovered in previous sections of this paper across the other sites of analysis.

I would also like to analyze a more traditional example of practice in the form of a project from the course, the Daily Design Journal. This is an introductory project learners begin at the start of the course, and the primary task is to document 14 objects learners encounter in their daily life. Learners are instructed to note the shape, form, materials, dimensions, functionality, and how those factors relate to the way they interact with the objects they choose. This activity is once again driven by the instructor, who sets the criteria for what matters and should be included in the final product, and also sets the tone for the reflective practice that follows the sketching portion where students respond to a series of questions about the three sketches of their choice. This activity doesn't illuminate much with regard to representation within the course, but it does demonstrate another opportunity for learners to evaluate disciplinary values and practice those values by learning to think like a designer by examining designed objects and reflecting on how their designs connect with their functionality.

Discussion

I recognize this analysis as presented here is only the beginning of what will hopefully become a much more in-depth, refined process. I also hope this documentation serves as a starting point for re-evaluating our instructional design processes from the beginning of a course design, rather than retroactively considering issues of power, identity, inclusion, etc. When reflecting on whether the framework uncovered previously hidden problematic areas, it seems more like this version of the framework offered a way to articulate those issues. For instance, a theme that became painfully clear in this course, and likely would for many in my portfolio, is that there is only one subject matter expert typically present in the development, and that leads to issues of power and representation that have been outlined above. While many may have assumed this would be a problem, I did not imagine the breadth and depth of the influences power can have on all aspects of the course. I went into this work assuming the critique activity was an opportunity for learners to have some power in the course, but in reality, my reflections within the other sites of analysis made it abundantly clear that this activity is not at all balanced due the instructor-imposed guidelines on interactions.

Another valuable outcome of this effort is disrupting the notion that I could break each piece of the course apart and analyze them individually. I quickly realized that this framework, at least as I interpreted it, leads to a lot of overlap and influences other parts of the framework.

For instance, the course procedures and policies dictate a lot of the social relations in the course. Participation grades in particular mean there has to be a designed interaction, and similarly, that need affects the practices that are designed into a course, which in turn influences the way learning is organized. All of these factors are constrained by the tools available. The choices made about course designs early in the process and the beliefs held by instructors and designers have a waterfall effect that touches nearly every experience designed into a course. This leads to the power problem I referred to that continually resurfaces throughout each of my chosen sites of analysis. It is the decision-making of the person or people in power that influences who is represented and what disciplinary values are prioritized, and how learners may develop new identities within these experiences. As a result, this discovery calls for further reflection around the notion of power and whether it should be positioned differently in the framework, or addressed in an entirely different way. In addition, the influence of power in our work leads to the question of what instructional designers can do to shine a light on this issue.

Reflecting on the critical focus of disciplinary values and identity development, I wonder now if these are two different categories. To elaborate further, alignment of disciplinary values is good from an instructional design perspective, but that doesn't necessarily mean the values themselves are inherently good. Those responsible for the learning design need to determine together if the values represent ethical outcomes for all, or if the values conflict. Additionally, learners also need to decide for themselves if the values they interpret are in alignment with their own personal values, regardless of the ethics in the disciplinary values. Further research into how students interpret disciplinary values, as well as how disciplinary experts interpret disciplinary values, could help this part of the analysis. Perhaps instructional designers can design opportunities for students to examine their personal values against the perceived values of their chosen area of study throughout their college careers.

After this initial trial, I also wonder if there is a better way to consider questions of representation in this framework? Looking at who is present in the referenced materials was productive, but I struggled to identify any other instances of representation in the practices, tools, or social relations within the course. There are undoubtedly different ways to think about representation that I have overlooked in this work as a result of my privileged way of experiencing education, and exploring research around heterogeneity in learning might benefit the next iteration of this framework. This gap may also require further consideration of stakeholders and how they think about ideas of representation and identity within the discipline, and whether the course design meets those expectations.

This work is limited in some ways. While I hope the framework will help drive critical analysis of existing learning designs across all disciplines, I am biased by my experience as an instructional designer who has primarily worked in professionalized disciplines. The three themes or critical focuses I outlined at the beginning of this paper may not be as functional for learning designs in fields with less direct applications of skills and knowledge. Another bias that may have unintentionally impacted this framework is my experience in online learning design, which is much deeper than my experience in traditional residential learning design.

In addition, this framework was influenced only by the critical scholarship that I have encountered. Undoubtedly there are other authors and papers unknown to me whose work may strengthen or contradict this framework. I hope this framework sparks conversation and collaboration around iterations for future use to evaluate other courses in my portfolio, and hopefully with instructional designers from other disciplines interested in this line of work. As I emphasized above, when one individual holds all the power, systemic problem areas remain undiscovered. The experiences of other professionals can only strengthen this effort.

Conclusions: The Instructional Designer's Role in Critical Instructional Design

Near the beginning of this paper, I cited Barab et al., who discuss the role instructional designers have in educational designs and the responsibilities that should be carried with this role. The authors state that "designers should regard their work in terms of its impact not on a situation directly but, rather on how users transact with the work, with each other, and with their contexts" (Barab, et al., 2007, p. 296). As an instructional designer, exploring critical scholarship in the learning sciences has changed my perspective on what my role is when collaborating with instructors and other team members to develop learning designs. I have always viewed this role as one of advocacy; for pedagogy, for innovative and memorable learning experiences, and for learners in the margins. While that hasn't changed, this analytical exercise has shown that previous ideas of who is in the margins are not sufficient. Based on my analysis of this course alone, I already feel that

I should advocate for change in the process itself by ensuring there are voices other than mine and the instructor I'm working with involved when we are making these design choices. As I discussed previously, power relations are abundant and tangled within the typical instructional design process, and drawing attention to this problem should be the goal of all instructional designers.

To borrow once again from Collier (2020), a "small move" I can make now could be working with my team to remedy the problematic spots highlighted in this analysis, and encouraging my teammates to apply this process to their own portfolios. Sharing our discoveries could help our unit find ways to embed these critical focuses from the beginning of a new project in the future, and help us make a plan for revising our old designs. Additionally, finding ways to bring more people into the design process can be another "small move" to tackle in the present. Specifically, striving to include student voices in the instructional design process in ways that are authentic and meaningful for both parties will be an immediate focus in my professional work going forward. I can advocate for spreading out the power in the instructional design process, ask the right questions, and point to the literature. By adopting this critical stance, I can be one of those instructional designers who takes steps to "build transformative models of what could be" (Barab, et al., 2007, p. 264).

References

Barab, S., Dodge, T., Thomas, M. K., Jackson, C., & Tuzun, H. (2007). Our designs and the social agendas they carry. Journal of the Learning Sciences, 16(2), 263–305.

Collier, A. (2020). Inclusive Design and Design Justice: Strategies to Shape Our Classes and Communities. EDUCAUSE Review. https://er.educause.edu/articles/2020/10/inclusive-design-and-design-justice-strategies-to-shape-our-classes-and-communities.

Costanza-Chock, S. (2020). Design Justice: Community-led practices to build the worlds we need. Cambridge: MIT Press.

Esmonde, I. (2017). Power and sociocultural theories of learning. In I. Esmonde & A. N. Booker (Eds.), Power and privilege in the learning sciences: Critical and sociocultural theories of learning (pp. 6–27). New York, NY: Routledge.

Okun, T. (1999). White supremacy culture. White Supremacy Culture. https://www.whitesupremacyculture.info/

Taylor, K. H. (2018). The role of public education in place-remaking: From a retrospective walk through my hometown to a call to action. Cognition and Instruction, 36(3), 188–198.

Vakil, S. (2020). "I've always been scared that someday I'm going to sell out": Exploring the relationship between political identity and learning in computer science education. Cognition and Instruction, 38(2), 87-115.

Vossoughi, S. & Gutiérrez, K. (2017). Critical pedagogy and sociocultural theory. In I. Esmonde & A. N. Booker (Eds.), Power and privilege in the learning sciences: Critical and sociocultural theories of learning (pp. 139–161). New York, NY: Routledge.

Wenger, E. (1998). Communities of practice: Learning, meaning, and identity. Cambridge, MA: Cambridge University Press.

Winner, L. (1986). Do artifacts have politics? In The whale and the reactor (pp. 19-39). Chicago: University of Chicago Press. https://ebookcentral.proquest.com/lib/pensu/detail.action?docID=557593

Author Biographies

Victor Azuaje – *Mount Saint Mary College, NY*

Victor Azuaje is Professor of Hispanic Studies at Mount Saint Mary College, NY. He has a Master's Degree in Linguistics from the University of Delaware, a Master's Degree in Spanish and a Ph.D. in Spanish from Temple University. He won the 2020-2021 MSMC Teaching Innovation Award: "Review of his courses indicates thoughtful and deliberate planning and execution as well as broad effectiveness and a novel way of employing technology." His teaching practice has been guided by the idea that a hybrid classroom is the perfect setting for an engagement version of the Turing test or imitation game: a student asks questions to an in-person and an online instructor—or the online persona of a hybrid instructor—to find out who is less engaging. The online instructor wins if the student fails to identify her. When Prof. Azuaje is not playing this game, he reads a lot and writes a bit. He is the author of the books La crítica de la obra ausente, winner of the "Enrique Bernardo Nuñez" Biennial Essay Award, and Bajo la sombra de Azazel: Sacrificio, alegoría y conflicto social en Ramos Sucre, winner of the "José Antonio Ramos Sucre" Biennial Essay Award.

Maha Bali – *American University in Cairo*

Maha Bali is Professor of Practice at the Center for Learning and Teaching at the American University in Cairo. She has a PhD in Education from the University of Sheffield, UK. She is co-founder of virtuallyconnecting.org (a grassroots movement that challenges academic gatekeeping at conferences) and co-facilitator of Equity Unbound (an equity-focused, open, connected intercultural learning curriculum, which has also branched into academic community activities Continuity with Care and Socially Just Academia, and a collaboration with OneHE: Community-building Resources. Most recently, she co-organized the Mid-Year Festival (MYFest) via Equity Unbound, a way of re-imagining professional learning online as nourishing, equitable, emergent, communal, creative and agentic. She writes and speaks frequently about social justice, critical pedagogy, and open and online education. She blogs regularly at http://blog.mahabali.me and tweets @bali_maha

Martha Fay Burtis – *Plymouth State University*

Martha Fay Burtis, is the associate director and learning developer at the Open Learning and Teaching Collaborative at Plymouth State

University. In this role, she supports faculty with instructional design and pedagogical innovation. Prior to coming to PSU, she was the founding director of the Digital Knowledge Center at the University of Mary Washington. At UMW, she also administered faculty and student development projects, including the Online Learning Initiative and Domain of One's Own. She has particular expertise and interest in digital literacy and pedagogy; student-centered teaching and learning; and critical instructional design.

Autumm Caines – *University of Michigan – Dearborn*

Autumm Caines is a liminal space. You will find her somewhere between designer, educator, and technologist; at the time if this writing she is working as an instructional designer at the University of Michigan – Dearborn. Autumm was a first generation non-traditional college student who brings a unique lens to her professional work as an advocate for those who often do not get a chance to participate in higher education. Autumm's work challenges status quo ideas of educational design and techno-solutionism in education through a critical analysis which weaves together constructs of student agency, digital privacy, equity, and surveillance. For examples of Autumm's projects, publications, presentations, and teaching experiences see https://autumm.org/

Amy Collier – *Middlebury*

As the Associate Provost for Digital Learning at Middlebury, Amy Collier provides strategic vision and leadership for Middlebury to create and sustain a global learning community through the effective use of digital pedagogies and technologies. In addition to leading Middlebury's Office of Digital Learning and Inquiry, she works on critical issues in higher education's intersections with digital technologies, such as privacy and student data, and equity and inclusion. She is the co-founder of Higher Education After Surveillance, a global group of educators and scholars imagining and developing alternatives to problematic visibility & surveillance in higher education. This group recently developed a prototype for a Higher Education Surveillance Observatory to help explore the landscape of surveillance and privacy across the higher education sector, and a Data Stories project, which invites speculative stories about surveillance that re-shape our thinking on its role in higher education. She is also also a co-facilitator of the Design Justice Network's Instructional Design Working Group, which aims to help Instructional Designers explore and adopt design justice principles in their work. Amy blogs rarely at redpincushion.me and tweets irregularly at @amcollier

Clayton D. Colmon – *University of Pennsylvania*

Clayton Colmon is the Associate Director of Instructional Design for Penn Arts & Sciences Online Learning. His work grows out of interests in race, gender, sexuality, science fiction, and digital studies, as intersecting entry points for examining technology's impact on community-building efforts for Black folks and minoritized groups in digital environments. He earned a BA in English and Political Science from Rutgers University and holds a Ph.D. in English from the University of Delaware. Clay believes that interdependent, lifelong, learning is integral to any sustainable social system. As such, he supports efforts to design spaces for pluralistic, justice-driven, pedagogies that can help us all get and stay free. In addition to his instructional design and curriculum development work, he teaches courses on many things, including digital strategies and culture, Black speculative futures, and queer projects for digital publics. You can find him on Twitter @warmclay.

Robin DeRosa - *Plymouth State University*

Robin DeRosa is the Director of the Open Learning & Teaching Collaborative at Plymouth State University, part of the University System of New Hampshire. The Open CoLab is a dynamic, praxis-powered hub dedicated to innovative teaching and learning and a community-driven approach to academic professional development; the CoLab focuses on instructional design, open education, interdisciplinary learning, and increasing the public impact of the academy.

Daniela Gachago – *University of Cape Town*

Daniela is an Associate Professor at the Centre for Innovation for Learning and Teaching (CILT) at the University of Cape Town. Her research focuses on academic staff development for blended and online learning in higher education, with a particular focus on developing socially just course and curriculum designs drawing from co-creative approaches such as design thinking in contexts of high inequality. She completed a Masters in Adult Education at the University of Botswana and received a PHD from the School of Education at the University of Cape Town. She is a C1 rated researcher and 2022 Fulbright Scholar has published more than 50 peer-reviewed articles and book chapters.

She is the managing editor of CriSTaL, the journal for critical studies in teaching and learning in higher education. She blogs at http://danielagachago.blogspot.com and tweets under @dgachago17

Jan Hare – *University of British Columbia*

Jan Hare is an Anishinaabe scholar and educator from the M'Chigeeng First Nation, located in northern Ontario. As an Indigenous scholar, educator, and administrator, she has sought to transform education in ways that are inclusive of Indigenous knowledges, pedagogies, and languages.. Her research is concerned with improving educational outcomes for Indigenous learners by centering Indigenous knowledge systems within educational reform from early childhood education, K-12 schooling, through to post-secondary, She designed and developed the Massive Open Online Course (MOOC), Reconciliation Through Indigenous Education, which has been taken by over 60,000 participants world wide. Jan holds a Canada Research in Indigenous Pedagogies in the Faculty of Education at the University of British Columbia, where she is Professor and Dean of Education.

Jeni Henrickson - *Middlebury*

Jeni is an instructional designer with Middlebury, working with faculty, staff, and students across the Middlebury campuses. As a member of the Office and Digital Learning and Inquiry (DLINQ), I help design pedagogical approaches, tech integration, and meaningful learning experiences in digital spaces across the institution.

Hannah Hounsell – *Plymouth State University*

Hannah Hounsell is the Learning Advisor for the Open Learning & Teaching Collaborative at Plymouth State University. Hannah has spent five years working professionally in the field of public, higher education. Her work has intersected with many different spheres, including student services and advising, instructional design, faculty/ staff professional development, and administration and operations. Her current position combines student advising for an indivdualized major program with faculty professional development. Hannah's teaching work began in the field of secondary education when she was certified to teach English in grades 5-12 and pursued several long-term substitution roles. Her scholarship interests include critical/ traditional instructional design, student-led professional development, and peer-to-peer mentorship. In 2021, Hannah received a graduate certificate in

Instructional Design and graduated in 2022 with her Masters in Education in Currriculum and Instruction, which inspired interest in critiquing frameworks and methods for traditional instructional design.

Surita Jhangiani – *University of British Columbia*

Surita Jhangiani is an assistant professor in the Faculty of Education at the University of British Columbia, Point Grey Campus, which is situated on the traditional, ancestral, unceded territory of the Musqueam people. She primarily teaches in the Bachelor of Education program and also teaches upper level graduate courses in the area of Human Development, Learning and Culture. Her research interests include scholarship related to open education, alternative grading, pedagogy of care and mental health. Her work is informed by a post-colonial, diversity and gender lens.

Meryl Krieger – *University of Pennsylvania*

Meryl Krieger is Senior Learning Designer for Penn Arts & Sciences Online Learning at the University of Pennsylvania. She also co-leads the Digital Strategies & Culture program within the Penn Liberal & Professional Studies Online program and teaches interdisciplinary social sciences courses in that program. Her focus is on accessible and equitable course development and nontraditional student learners, and sees online learning as a tool for increasing access and opportunities for quality education to the widest possible audiences. Meryl earned her PhD in Folklore & Ethnomusicology from Indiana University Bloomington, which became a focal point for her developing research and teaching practices on digital technologies and the learning process.

Sarah Lohnes Watulak – *Middlebury*

As the Director of Digital Pedagogy and Media in the Office of Digital Learning and Inquiry (DLINQ) at Middlebury College, Sarah leads a team of instructional designers focused on creating digital learning opportunities and environments that support learner agency, equity, and critical engagement with the digital. She also leads DLINQ's Inclusive Design Studio, a hub for exploration and application of inclusive design and design justice practices across contexts. Sarah earned her Ed.D. from Teachers College, Columbia University in 2008, and before joining Middlebury, was an Associate Professor of Instructional Technology in the College of Education at Towson University. For more on Sarah's recent research and writing: http://sarahlw.middcreate.net/

and @sarahlw6 on Twitter.

Mary Mathis Burnett – *Arizona State University*

Mary Mathis Burnett is an instructor and Manager of Instructional Design and Inclusive Pedagogy in Watts College of Arizona State University, has more than 15 years experience in instructional design, and holds an Ed.D. from Mary Lou Fulton Teachers College at ASU. Her lived experiences activate her passion to increase awareness of her own biases, those unavoidably acculturated from society and the more socially just ones created in the vacuum the old ones leave behind. Her work joins the efforts and experiences of others in the field moving to disrupt the apparent power dynamics in higher education, advocating for equity and access, and reducing harm done by systems, structures, and the status quo to those with racialized or marginalized identities. You can find her on Twitter (@burnettical) and at https://critpedframework.com

Cynthia Nicol – *University of British Columbia, Canada*

Cynthia Nicol is Professor in the Department of Curriculum and Pedagogy and holds the David F. Robitaille Professorship in Mathematics and Science Education. Cynthia works alongside rural, remote, Indigenous and other communities to explore new strategies for supporting teachers in designing and implementing culturally responsive mathematics education and STEM (Science, Technology, Engineering, and Mathematics) education. Her current projects include researching ways to support all learners, including teachers, to connect math, community and culture, with land-based pedagogies and Indigenous storywork. Cynthia is committed to building capacity in communities for learners to use mathematics and STEM, to not only interpret the world, but also change it. You can find her work at the Indigenous Math Education Network: https://indigenous.mathnetwork.educ.ubc.ca/ and the David Robitaille International STEM Network: https://dfr.stemnetwork.educ.ubc.ca/

Nicola Pallitt – *Rhodes University*

Nicola Palitt is an Educational Technology Specialist at senior lecturer level in the Centre for Higher Education Research, Teaching and Learning (CHERTL) at Rhodes University in South Africa. She holds a PhD in Media Studies from the University of Cape Town. Nicola's research interests include online learning design, networked professional development and online supervision. She supervises postgraduate students, co-teaches formal courses in Higher Education and co-facilitates professional development opportunities for lecturers in various settings. She is part of the e/merge Africa team, an online professional development network for educational technology researchers and practitioners in African higher education. She is also involved in the leadership of the Culture, Learning and Technology (CLT) division of the Association for Educational Communications and Technology (AECT). Read more about Nicola here or follow her on Twitter @nicolapallitt

Nicola Parkin – *Flinders University*

Nicola Parkin likes questions, enigmas, and human depths. She is a higher education practitioner who locates her work in the difficult spaces between teaching, learning, professional practice, philosophy, design, and creative practices. Nicola's educative commitments have been forged over many decades of work in community cultural development and adult community education programs – however, it is the wild untamed inner forces of learning and becoming in each individual that most delight and intrigue her. Methodologically, Nicola leans in to the existential, the hermeneutical, the phenomenological. Her research is about finding meaningful ways to see, say and save the deeper undercurrents of our educative work, and how we might purposefully and profitably bring these qualitative riches into our everyday practices. She has worked at Flinders University in South Australia for many years supporting program and learning design. You can find a bit of Nicola's musings on bebetween on WordPress.

LeAnne Petherick – *University of British Columbia*

Dr. LeAnne Petherick's areas of special interest revolve around the meaning of production in physical education and health education, including curriculum design, delivery, and uptake. Her research focuses on the development and translation of ideas about the body, children and youth, teachers and health professionals, and contemporary health culture. Furthermore, she examines the embodied experiences associated with these ideas as they are mobilized in educational contexts. Within this area of special interest, her research centres around two primary themes: 1) social justice into physical and health education; 2) community-based Indigenous physical cultures (i.e., sport, physical activity, land-based activities) and place-based learning.

Jerod Quinn – *Wake Forest University*

Jerod Quinn is an instructional designer at Wake Forest University in the Office of Online Education. He's been an ID for well over a decade helping instructors create online classes they are excited to teach and that learners are excited to take. He has a Master's Degree in Learning Systems Design and Development and is a dissertation away from a PhD in Educational Psychology with an emphasis in Quantitative Measurement from the University of Missouri. His work bubbles out of the tension of holding the pragmatic (what can actually be done) in one hand while holding the ideal (what should be done) in the other. His first book, The Learner Centered Instructional Designer, came out in 2021 through Stylus Publishing. You can find him tweeting about teaching, learning, instructional design, guitars, and nature at @jerodq.

Johanna Sam – *University of British Columbia*

Hunelhyad? Sid Dr. Johanna Sam sets'edinh. Sid Tŝilhqot'in xaghiyah. Sid Musqueam nen ŝidah as. My name is Dr. Johanna Sam. She is a proud citizen of Tŝilhqot'in Nation. She lives and works on the traditional lands of the Musqueam People. She is an Assistant Professor in the Faculty of Education at the University of British Columbia. Realizing the importance of a strength-based approach, she is involved in creating youth-friendly educational and mental health resources. Her research explores the relationships among cyber-aggression, resiliency, academic achievement, and wellness. Her research and teaching not only utilizes digital technology, but approaches those digital tools from Indigenous perspectives.

Natalie Shaw – *NHL Stenden University of Applied Science*

Natalie enthusiastically works as a teacher educator after a career in international education, both in a classroom teacher and in a leadership role. She currently teaches on the ITEps course, a unique B.Ed. programme located on a small campus in Meppel, the Netherlands. At ITEps, student-teachers are trained specifically for international education, with a big focus on teaching in diverse classrooms and exploring global citizenship education. Natalie completed a university degree in Special Needs Education from Cologne University and went on to obtain a PGCE in Primary teaching with Early Childhood focus at Liverpool John Moores university. At the same time, she ventured to Asia regularly, volunteering in community-based rehabilitation in Thailand and an outreach school for young children in India. After teaching positions in China and Cambodia, Natalie returned to Germany to teach in Berlin and completed both a Masters degree in Children's Rights and Childhood Studies (at Freie Universität Berlin) as well as a related PhD (at the University of Kassel). Her educational adventure continued when making the transition into teacher education, and she has not looked back since. At ITEps, she teaches Early Childhood Education, Educational Studies, and Research. In her free time Natalie can be found doing diy on her houseboat, sewing, crocheting, painting or crafting, and doing sports of all kinds.

Jesse Stommel

Jesse Stommel is currently a faculty member in the Writing Program at University of Denver. He is also co-founder of Hybrid Pedagogy: the journal of critical digital pedagogy and Digital Pedagogy Lab (2015-2021). He has a PhD from University of Colorado Boulder. He is co-author of An Urgency of Teachers: the Work of Critical Digital Pedagogy.

Jesse is a documentary filmmaker and teaches courses about pedagogy, film, digital studies, and composition. Jesse experiments relentlessly with learning interfaces, both digital and analog, and his research focuses on higher education pedagogy, critical digital pedagogy, and assessment. He's got a rascal pup, Emily, a clever cat, Loki, and a badass daughter, Hazel. He's online at jessestommel.com and on Twitter @Jessifer.

Katrina Wehr – *Pennsylvania State University*

Katrina Wehr is a U.S.-based learning experience designer currently working for WGU Labs and wrapping up a Ph.D. in Learning, Design, and Technology at Penn State University after attaining her M.Ed. in instructional technology from Kutztown University of Pennsylvania in 2013. Her experience serving the educational needs of leaners spans both K-12 and higher education. A beneficiary of online learning herself, Katrina is inspired by the achievements of learners who otherwise could not access higher education without flexible distance learning opportunities. Lately, her professional priorities are focused on providing individual contributors in education settings with knowledge and tools to take actionable changes within their organizations towards ethical, justice-oriented learning design. She spends her personal time walking her dogs, riding her bike, and spending time with family. Katrina respectfully acknowledges she is situated on Lenapehoking, the ancestral homeland of the Unami Lenape.

www.ingramcontent.com/pod-product-compliance
Lightning Source LLC
Chambersburg PA
CBHW070642160426
43194CB00009B/1552